Studies in Brain and Mind

Volume 24

The series Studies in Brain and Mind provides a forum for philosophers and neuroscientists to discuss theoretical, foundational, methodological, and ethical aspects of neuroscience. It covers the following areas:

- Philosophy of Mind
- Philosophy of Neuroscience
- Philosophy of Psychology
- Philosophy of Psychiatry and Psychopathology
- Neurophilosophy
- Neuroethics

The series aims for a high level of clarity, rigor, novelty, and scientific competence. Book proposals and complete manuscripts of 200 or more pages are welcome. Original monographs will be peer reviewed. Edited volumes and conference proceedings will be considered provided that the chapters are individually refereed.

This book series is indexed in SCOPUS.

Initial proposals can be sent to the Editor-in-Chief, prof. Gualtiero Piccinini, at piccininig@umsl.edu. Proposals should include:

- A short synopsis of the work or the introduction chapter
- The proposed Table of Contents
- The CV of the lead author(s)
- If available: one sample chapter

We aim to make a first decision within 1 month of submission. In case of a positive first decision the work will be provisionally contracted: the final decision about publication will depend upon the result of the anonymous peer review of the complete manuscript. We aim to have the complete work peer-reviewed within 3 months of submission.

For more information, please contact the Series Editor at piccininig@umsl.edu.

Camden Alexander McKenna

Constructing the Present

An Investigation into Time-Consciousness

Springer

Camden Alexander McKenna
Department of Philosophy
University of Edinburgh
Edinburgh, UK

ISSN 1573-4536 ISSN 2468-399X (electronic)
Studies in Brain and Mind
ISBN 978-3-031-89570-8 ISBN 978-3-031-89571-5 (eBook)
https://doi.org/10.1007/978-3-031-89571-5

This Springer imprint is published by the registered company Springer Nature Switzerland AG
The registered company address is: Gewerbestrasse 11, 6330 Cham, Switzerland

To my family, my friends, my teachers, and my students
and to all those suffering from chronic pain—I wish for time to be your salvation as well as your enemy, as it has been mine.

Acknowledgments

Had it not been for my colleague Kathryn Nave first having the idea to turn her own research into a book, now titled *A Drive to Survive: The Free Energy Principle and the Meaning of Life* (Nave 2025), and had it not been for my PhD supervisor, Alistair Isaac, believing that I could do likewise, this book would certainly not exist. My deepest appreciation therefore goes out to Kate for showing me the door and to Alistair for encouraging me to walk through it.

This book is the result of a substantial reworking of my PhD dissertation at the University of Edinburgh and therefore owes a huge amount to the always insightful consideration of my primary supervisor Alistair Isaac, as well as my secondary supervisors, Alasdair Richmond and Keith Wilson. Mazviita Chirimuuta, who served as internal examiner for my doctoral dissertation, and Craig Callender, who served as my external examiner, also provided crucial encouragement in pursuing this project beyond the PhD. I also want to give special thanks to Dave Ward, who supervised my earlier MSc dissertation, which was the seed from which my interest in this topic has grown.

Throughout my time at the University of Edinburgh there was great excitement about predictive processing and 4E cognition (embodied, embedded, enactive, and extended), especially as these ideas were expressed by Andy Clark. It was after reading Clark's *Surfing Uncertainty* (2016) and Jakob Hohwy's *The Predictive Mind* (2013) in Alistair Isaac's graduate seminar and especially following Jakob Hohwy's visit to Edinburgh in 2017 that I realized there were unrecognized implications of the predictive processing view for how we might understand time-consciousness. Even then, I was beginning to think that we construct the present *en passant* out of the operation of the predictive engine of the mind, but I had not yet given the idea comprehensive thought. I then developed the bulk of the ideas in this book during my doctoral studies, and I must thank everyone at the University of Edinburgh who formed the unique milieu during that period, which influenced me and resonated so well with my own thinking.

I would like to thank everyone who participated in the Time and Perception Reading Group, which I started in 2018 to discuss some of the key background texts relevant to this book with like-minded people working on similar issues. I am grateful for constructive conversations with Keith Wilson, Alisa Mandrigin, Cain Todd, Kate Nave, Derek Maccoll, Martin Miragoli, Albinas Plesnys, Preston Haselton-Parke, Olivia Coombes, and anyone else who came along for however long. I'd also like to thank Victoria Turner for inviting me to respond to John Dupré's 2023 Gifford Lectures on the philosophy of biology (*A Process Perspective on Human Life*), which turned out to be a source of new and unexpected inspiration for me.

Additionally, I am grateful to anyone who has provided feedback, comments, questions, or raised eyebrows during my presentations on the topics I have addressed in this book. Specifically, I'd like to thank Jack Shardlow for his helpful response to my presentation of the ideas from Chap. 3 at MindGrad 2021 as well as Adrian Bardon and Suzuka Komatsu for further correspondence about my presentation and papers. Suzuka deserves special mention for providing the translation of the poem in the preface, for providing feedback on a number of sections, and for being incredibly supportive during a particularly difficult time in my life.

Some of the ideas that became Chap. 3 were also presented at the Australasian Postgraduate Philosophy Conference in 2020 and the Italian Society for Analytic Philosophy's Annual Conference in 2021, and I am grateful to the audiences at both for their feedback. I presented some of the material that became Chap. 6 at the 2022 Salzburg Conference for Young Analytic Philosophy and am grateful especially for the enthusiasm that was shown for that presentation. Aspects of Chap. 6 were also presented to the Active Minds Group at the University of Edinburgh in 2022. Some of the material that became Chap. 7 was presented at a talk for the Mind and Cognition Group at the University of Edinburgh in 2017 as well as at the "Worlding the Brain" Conference at the Aarhus Institute for Advanced Studies in Denmark in 2018, and I'd like to thank the audience at the latter, especially Julian Kiverstein and Uffe Schjødt, for their thoughtful and supportive responses.

This book has benefitted enormously from the careful attention of several anonymous peer reviewers whose uncompensated and yet critically important efforts make the academic world go round. While the chapters with previously published versions have of course benefitted from the eyes of several peer reviewers, I would particularly like to thank the reviewer entrusted by Springer with the entire book, who carried out this monumental task exceptionally quickly, while still providing insightful and charitable comments.

Special thanks go out to Ravi Thakral, Ian Mason, George Deane, and Giulia Burlando for reading over and providing feedback on various sections. Ravi Thakral deserves another special mention for assigning the published version of Chap. 3 as one of the core readings in his metaphysics courses at the University of Sussex and later at the University of Nevada, where I trust it generated a spirited debate about heptapod consciousness. Ian Mason was kind enough to read through an early draft cover to cover and provide a scholarly perspective from outside the discipline of philosophy. I'd also like to thank Kyle Rattray for helping to proofread Chap. 5 prior to its initial publication, when he certainly had better things to do with his time. Alex Norcott and Peter Morrison also deserve acknowledgement for helping to brainstorm a compelling title over pancakes at Sir Walter Scott's house.

My gratitude also goes out to all the students from the Philosophy of Time course who attended my lecture on Temporal Experience and the Specious Present in Spring 2024, especially those that provided feedback. In addition, I'd like to thank the students at the PPLS Skills Centre in Edinburgh who have shown interest in my research, especially Julia Ogrydziak, who read and discussed my work while writing her MSc dissertation.

Finally, this book is above all dedicated to my family for staying the course with me and for keeping me sane and healthy through good times and, more recently, bad. It is nearly impossible when reading a dry academic text to discern the realities of the writing process, but rest assured, the saga unfolding behind the scenes as this book was being written is more than worthy of its own book-length treatment.

Preface

Ancient pond
Frog jumps in
Sound of water

—Matsuo Bashō [1686][1]

There is nothing more familiar to us and yet more mysterious than what we call the present. It is always there in the background of our lives, even if so many of us feel out of touch with it. We may long to capture it and freeze it, so it doesn't slip away, just as Matsuo Bashō seems to have done in his iconic poem. And yet we also realize that our attempts are fruitless. Our present always departs as fast as it arrives. For Bashō, just as for ourselves, the present is forever ephemeral.

This book is concerned with time as experienced by living beings. More specifically, it is concerned with what I take to be the most basic feature of subjective time: the experiential present. The book sets out to answer two questions regarding the experiential present. These are (1) "What is it like?" and (2) "How does it come to be like that?". I believe that in pursuing the study of phenomenology—the careful analysis of our experience—we can learn some things about what our present really is, clearing away some of the philosophical confusion surrounding this most important aspect of our existence in this world. At the same time, I think contemporary cognitive science, psychology, and neuroscience have a lot to teach us about how to understand the present and our own living, experiential time. By investigating our minds from an interdisciplinary perspective, we stand to gain a more complete and coherent understanding of our human condition, which is fundamentally and inevitably temporal.

My goal with this book is to transform the common and scholarly understanding of subjective time. I will argue that *our* present, as opposed to what physics may call the present, is not just granted to us by the universe. It is not a thing or property out there in the world that we simply pick up on, in the way some might think we perceive the shape of a room. Instead, I argue that we *construct the present* on an

[1] Thank you to Suzuka Komatsu for this translation.

ongoing basis as we construct our own subjective reality and as a function of the very same mechanisms that underlie consciousness generally.

In the course of this investigation, I make a number of claims. I regard none of these as the final word on the subject. But I hope to provide solid enough reasons for these claims that others may be inspired by them when they do their own thinking on this often neglected, yet fundamental subject. I will not shy away from addressing empirical issues and recruiting evidence from the sciences where it is relevant. In general, it seems to me that very little about the mind and how it works is revealed *a priori*. I therefore consider both phenomenological analysis and the sciences of the mind to be appropriate and entirely compatible methodologies with which to advance our understanding of experiential time.

I began thinking about this topic in the reverse order to how the text proceeds. By this I mean that it was through an interest in predictive processing and cognitive science, topics which are touched on towards the end of the book, that I entered the literature on temporal experience. I think this is an atypical direction of travel for philosophers writing on time and phenomenology of time, who I suspect typically start from an antecedent interest in metaphysics or pure phenomenology. However, I think this unusual trajectory goes a long way towards explaining my unorthodox perspective and interests. To put a fine point on it, I think it is probably the case that readers approaching this book from the field of philosophy of time will likely not be concerned with the same issues as I am here, while those who are approaching this book from the perspective of the philosophy of mind may find it resonates with them more. Rather than concerning myself with questions like "is time real?", "is temporal passage real or illusory?", "do we live in a four-dimensional block universe?", etc., I am really much more focused on the mind, in the tradition of those who believe that our own consciousness is the central mystery of our lives.

I am concerned first and foremost, in other words, with the nature of experience: how it is, and why it is as it is. Because of its centrality in our subjective experience, time is a topic that would seem to deserve a great deal of attention in the philosophy of mind and cognitive science but is instead most often considered through the distorting lens of metaphysics. Like David Hume, with whom I have so often shared a location in space but not in time, I tend to see metaphysics as an enterprise built on shaky foundations. In contrast, that which is given in our experience, which is both utterly perplexing and entirely taken for granted, seems to call out in the first place for philosophical attention. The rest of this book will attempt to answer that call by analyzing our experience and arguing that what is "given" is really the result of an activity our minds are continuously engaged in: constructing the present.

Edinburgh, UK Camden Alexander McKenna

A Note on the Text[2]

> Books, even books about history, are solid things, susceptible to the gnawing ants of time.[3]—Mazviita Chirimuuta (2024, x)

Books are indeed solid things in a world that is not. It is a consequence of the pace of academic life today that even before finishing a manuscript, there will already be a multitude of relevant new journal articles, new theories, and even new figures in the field, that perhaps some people will expect to have been covered. This is without even mentioning all the relevant work of the past that the author, for one reason or another, fails to include. To this I can only say my thoughts are necessarily partial and incomplete, and I must encourage my few interested readers to take this book as an invitation to fill in the gaps, and not as a declaration that I have closed them.

---

[2] A further note: it occurred to me toward the end of the production process in 2025 that the use of artificial intelligence has now become so widespread that a book written without any resort to AI consultation might already seem anachronistic. Yet I think for humanists in particular, it is important that we remain anachronistic in this way, and so I have remained.

[3] Given that Mazviita has personally witnessed the development of this monograph over time, I thought it especially appropriate to bring attention to this quote from the preface of her recent book, *The Brain Abstracted* (Chirimuuta, 2024).

Contents

Chapter 1
Introduction: The Perennial Problem

Sixteen hundred years ago, the North African Bishop Augustine reflected on the nature of time: "What then is time? If no one asks me, I know what it is. If I wish to explain it to him who asks, I do not know," (*Confessions* XI, xiv). Today this question remains as vexing as ever. I will argue in the course of this book, however, that progress towards answering Augustine's question is possible, so long as we approach the subject with care. One way to answer the question is by attempting to characterize time as external to and apart from ourselves. Another way is to look inward and investigate what time is like for *us*. It is this second pathway that will be the focus here, but the distinction has not always been clear. In part, the confusion is semantic. The problem is that the word "time" is ambiguous between several different concepts, many of which, but not all, refer to aspects of our own multifaceted experience. Conceptual analysis is therefore necessary to avoid misunderstandings and precisely delimit our subject of investigation.

We can begin by cordoning off the physical concept of time. This concept is itself continuously evolving. The notion of a physical absolute time, that is, a universal temporal reference frame, was commonplace from Aristotle's era through Newton's and beyond. Following Einstein, relativistic physics rejected the universal temporal reference frame. Instead, it came to be thought that space and time constitute a unified manifold: spacetime. In the twentieth century, how "time" or "spacetime" operate under quantum mechanics became its own puzzling question, along with how to reconcile quantum phenomena and phenomena on a macro scale that appear to behave relativistically. In each case, however, the time of physics is conceptualized as a feature of the universe we find ourselves in. This sort of time is often called "objective time" or "clock time" and refers to whatever it is that clocks measure, with the details deliberately left open to be filled in. Clock time remains a matter of great concern and considerable disagreement in physics and metaphysics, but this book will be turning the focus closer to home.

More immediately concerning to us as living subjects is the time of our *experience*, which also grounds many of our basic intuitions about time beyond ourselves.

C. A. McKenna, *Constructing the Present*, Studies in Brain and Mind 24,
https://doi.org/10.1007/978-3-031-89571-5_1

This is our first major conceptual distinction: "clock time" vs "subjective time". Augustine's question, however, applies equally well to either concept. What do we mean by "subjective time"? It seems all we know is that there is at least *something* that we notice upon introspection and then call "time".

One thing we call "time" is our abstract conception of time itself. For instance, one might conceptualize time in their life as having certain qualities—receding faster and faster as we age, for instance, or presenting us with a vast unknown ahead and the "dry bones of the past" behind, as Emerson (1836, p. 5) put it. But these broad conceptions seem to rest upon a firm and immediate phenomenological foundation. Our immediate temporal experience includes, among other things, our memory and imagination. We recall events we have already experienced in the present and we project future eventualities and can simulate them now. Both of these mental activities have a distinctive phenomenology—a certain way they are like to the subject.

However, neither memory nor imagination seem entirely essential to temporal experience. It would seem there are more basic phenomena still. The bare succession of experiential events that lends to our experience that familiar feeling of rolling inexorably into the future and away from the past—an experience itself beyond the ability of metaphor to effectively capture[1]—seems to be at the heart of what is meant by "time." This fundamental way that we experience the world, namely, in time, on a constant basis, has been termed "time-consciousness" by Edmund Husserl (1917), which is an apt label.[2] Time-consciousness is not necessarily consciousness *of* time, nor is it "time itself" in the metaphysical sense, nor does it exhaust the possible mental activities that have to do with time. Rather, time-consciousness refers to the way that our experiences are experienced. It is the way we experience the world right now. It is the felt, experienced, subjective present.

We have disentangled multiple senses of "time" and landed at "time-consciousness" as most fundamental to our own experience. Still, it seems Augustine's question is just as reasonable to ask as ever. What, then, is *time-consciousness*? In other words, how are we to understand this fundamentally temporal experience of ours? Historically, there have been many attempts to answer this

[1] The metaphors we use to elucidate experiential time are unusual in that they are themselves inscrutable without appeal to the temporal processes they are meant to describe. For instance, the only way to understand the metaphor of the "stream" of consciousness is through the stream of consciousness itself. Accordingly, it is unclear to what degree the metaphor serves to explain the phenomenon, when it is itself underwritten by the phenomenon it is meant to explain (for a lengthier discussion of this point see Merleau-Ponty, 1945, pp. 477–479). This is quite different than ordinary cases of analogical explanation (e.g., "a family is like a tree") where the *relata* are independent. Nonetheless, metaphors like "stream of consciousness" give us words where previously we might have had none.

[2] I use the term time-consciousness due to its historical pedigree. Others, most notably Dainton (2023), have used the term "temporal consciousness," but time-consciousness seems to me more prevalent, and can be found in contemporary psychology and neuroscience literature as well. As in philosophy, there the term usually refers to the extended experiential present, however it is to be understood (see e.g., Kent & Wittmann, 2021, who are careful to distinguish time-consciousness from "timing," as the latter often turns out to be what psychologists are directly studying).

question from a variety of angles, whether philosophical, psychological, neuroscientific, anthropological, sociological, theological, etc. However, strong interest in the analysis of the phenomenology of time-consciousness *per se* has only really come into its own over the last century and a half or so. Despite increasing philosophical interest in the nature of our subjective experience, the metaphysics of time continues to attract much more attention. When philosophers do consider temporal experience, it is often from the perspective of metaphysics, with experience serving as a constraint on the way the world might be. However, given that temporal experience is so central to our existence—even, as I will argue, an enabler of this kind of existence—the study of temporal experience in its own right warrants much more attention than it has historically received. It is my hope that this book contributes to the renewal of interest in the philosophy of subjective time and the repositioning of the questions surrounding the topic towards the center of philosophical discourse.

We can see the many attempts to analyze the experience of time as occupying different levels of abstraction. Some of these levels are "high up" in the sense of embodying highly abstract characterizations of the structure of our phenomenology. In other words, they give some account of our temporal experience beyond just saying that such a thing exists. These analyses are usually geared towards answering puzzles about our experience or providing an explanation for certain phenomena in our experience. We can call this the macro-phenomenological level of analysis. Examples of this sort of analysis would be the breaking down of our temporal experience into a successive structure or observing that we experience the present as an extended duration (the so-called *specious present*).

The macro-phenomenological level stands in contrast to the "micro-phenomenological" level of analysis (Dainton, 2010), which represents a more fine-grained description of the constituents of our temporal experience. This level has probably exercised philosophers more than any other over the last century or so. It is at this level that determinations about the internal composition of experience are made, which may not be manifest to introspection, but which may nonetheless turn out to be required for experience to be as it is. The influential "retentional" phenomenological analysis of Edmund Husserl (1917), which claimed experience exhibits an invariant tripartite temporal structure, falls largely on this level (much more on the details of that account later).

Furthermore, there are also implementation-cognizant accounts, often coming from the sciences of mind, which themselves occupy varying levels of abstraction. Accounts can be given at the functional level, for instance concerning information processing, perhaps outlining a possible mental computational strategy that is meant to result in a particular kind of temporal experience. Accounts can also be given of the alleged neurophysiological correlates of time-consciousness. Accounts given at these various levels are not necessarily exclusive or in competition, and, ideally, should all hang together coherently to give a complete picture of how temporal experience works for us.

This book proposes plausible phenomenological analyses of time-consciousness on both the macro and micro levels, which not only solve longstanding conceptual problems, but also dovetail nicely with an information processing implementation that I will introduce here as well. The overall picture should be a coherent account of temporal experience on several levels of analysis that accommodates relevant phenomenology, with advantages over rival explanations and practical implementation-level implications.

I will be taking a broadly adverbialist position on time-consciousness as an overarching attitude.[3] Temporal adverbialism, as I define it with regards to subjective time, is the idea that time itself is not experienced but rather we experience the world in a temporal way. In keeping with this adverbialism, I have tried to avoid the phrase "time perception," which may lead readers to mistakenly view time as an object of perception like any other. Unlike adverbialism about other aspects of perception, the position that time is not perceived *per se* but rather we perceive temporally has been fairly common at least since the time of Kant, even if it is often implicit. As recognition dawns that there is "no dedicated sensory system for time" that picks up on particular properties of the world (Seth, 2021, p. 134), I believe this sort of adverbialism should become increasingly attractive. This view opposes the fruitless search for a "sense of time," including the traditional search in psychology for "internal clocks" that ground basic phenomenology. While biological rhythms are certainly regulated by circadian pacemakers, and it may be that temporal *judgments* rely on certain neural or cognitive timing mechanisms, adverbialism about subjective time forecloses a further explanatory role for internal clocks with regard to time-consciousness, in part because the meaningfulness of a clock's content already presupposes a temporally structured subjective experience of the clock itself.

My proposal endorses the specious present and the centrality of succession to temporal experience, as well as the apparently felt passage of time for the experiencer. These are claims about macro-phenomenology. On a micro-phenomenological level, the proposal combines aspects of the positions known as retentionalism and extensionalism, while attempting to shed their respective disadvantages. I call this view ERA, short for the "extensional-retentional analysis" of temporal phenomenology. Briefly, I am defining extensionalism here as the proposition that experience is extended over an actual duration of clock time, and we experience an extended present as such in virtue of this extension. Retentionalism, on the other hand, typically maintains that the actual temporal extension of our experience is explanatorily irrelevant, with the character of our experiences depending instead on the content of "intentions" that may be oriented towards different times (having content *as of* the just past, etc.), which themselves can have any particular ordering in time and are often schematized as existing together simultaneously and instantaneously, though representing different times to the subject.

[3] My proposed adverbialism about temporal experience and adverbialism as a theory in the metaphysics of time (e.g., Lowe, 1987) are entirely independent positions. The adverbialism adopted here has more in common with forms of adverbialism proposed in other domains of experience like color vision (see, e.g., Chirimuuta, 2017), whereby one sees "blue-ly" instead of seeing the color blue. Similarly, the adverbialism about time that I endorse stresses that we experience temporally and do not experience time. Like other forms of adverbialism, it is the manner in which something is done (which adverbs describe), rather than what is done, that is paramount.

ERA suggests the actual extension in clock time of the intentional entities posited by retentionalism *is* explanatorily relevant to the experiences we have. Specifically, the actual temporal extension forms the basis for the duration of the specious present, which I will argue would otherwise be inexplicable. Such a position fits nicely with currently prominent "predictive" approaches towards mind and brain function. These approaches provide information processing level accounts that could underpin the phenomenology analyzed above in a coherent way. I will accordingly offer a sketch of what one such predictive account might look like towards the end of the book, which I call TIER: Temporality as Iterative Expectation Revision. This account understands subjective temporality to be the upshot of the iterative expectation revision performed by an always-active, predictive cognitive system. On this view, in generating experience itself we are always engaged in the continuous activity of constructing our own experiential present.

While I believe the context, motivations, and ancillary commitments discussed throughout the book are critical for a full understanding of the positive proposal, those looking to skip straight to the most provocative claims will find them in Chaps. 6 and 7. These chapters introduce ERA, i.e., the phenomenological analysis, and TIER, its proposed cognitive implementation. Otherwise, I recommend following the chapters in sequential order, especially for those readers who are not already familiar with the topic.

The plan of the book is as follows. **Chapter 2** addresses the question of what time-consciousness is in a more detailed but still pre-theoretical way. In other words, this chapter identifies the features of experience that a good theory of time-consciousness should accommodate and ideally explain, i.e., the *explananda*. In doing so, this chapter provides the raw material for the deeper phenomenological analysis found in later chapters.

Chapter 3 will look beyond the contingent features of our actual experience to the constraints that might exist on any possible experience.[4] More specifically, in this chapter I will argue that experiential succession is a requirement for agentive, subjective experience in this world. It is thus paramount that our theory of time consciousness makes this core feature fundamental and not an ad-hoc or "bolt-on" module such as an independent "sense of time."

Chapter 4 will review the high-level "classic" analyses of the phenomenology of temporal experience. These include the standard categories of extensional, retentional, and cinematic views. This chapter discusses the advantages and shortcomings of each and ultimately concludes that none of them on their own are satisfactory, thus pointing the way to a more sophisticated phenomenological analysis.

Chapter 5 will critically analyze a recent attempt to rehabilitate the cinematic view in the form of what's known as the "dynamic snapshot theory."[5] I will argue here these attempts are misguided and rest in large part on a mistaken interpretation

[4] A previous version of this chapter has been published under the title of "Agency and the Successive Structure of Time-Consciousness" (McKenna, 2023).

[5] A previous version of this chapter has been published under the title of "Don't Go Chasing Waterfalls: Motion Aftereffects and the Dynamic Snapshot Theory of Temporal Experience"

of a class of illusions known as motion aftereffects. With the snapshot theory in its various permutations in relative disrepute, we are left with "specious present" views as the most credible analyses of time-consciousness. I further argue here that identifying the mechanisms responsible for time-consciousness in localized neural processes is misguided and that instead we should seek explanations in system-wide features of the mechanisms responsible for consciousness generally.

Chapter 6 argues that a pluralistic explanatory strategy that applies different phenomenological analyses to different explanatory targets should be more effective than any of the classical views alone. In this chapter I present several positive arguments in favor of a view that combines the best elements of two of the classic "specious present" views, namely retentionalism and extensionalism. The specific view I propose is called the extensional-retentional analysis (ERA) of time-consciousness. I contend that it is capable of sidestepping many of the problems that confront the models it draws on, while retaining their explanatory power and fitting better with predictive approaches, as discussed in greater detail in Chap. 7. I intend Chaps. 6 and 7 to be read together for the fullest understanding of the overall position.

Chapter 7 argues that predictive approaches to the mind naturally dovetail with the kind of analysis of temporal experience described in the previous chapter. Specifically, this chapter proposes a model of "Temporality as Iterative Expectation Revision" (TIER), that aims to avoid some of the problems associated with other recent computational models that attempt to do justice to our temporal phenomenology. This chapter also provides further arguments for dispensing with the "internal clock" view and adopting an overall orientation of adverbialism, while also speculating on the biological and evolutionary origins of time-consciousness and suggesting that our particular cognitive and phenomenological setup may actually be inevitable given the physical characteristics of the world we inhabit.

Chapter 8 recaps the steps taken to arrive at each of the book's various proposals and addresses their significance in the larger philosophical and scientific context, with special consideration to the neural correlates of time-consciousness. Ultimately, I conclude that subjective time is the result of the ongoing processes that construct our experiential reality. The upshot is that, properly understood, time should not be seen as an object of perception but rather the way we experience the world.

References

Chirimuuta, M. (2017). *Outside color*. MIT Press.

Dainton, B. (2010). *Time and space* (2nd ed.). Acumen.

Dainton, B. (2023). Temporal consciousness. In E. N. Zalta and U. Nodelman (Eds.), *The Stanford Encyclopedia of Philosophy* (Spring 2023 Edition). https://plato.stanford.edu/archives/spr2023/entries/consciousness-temporal/.

(McKenna, 2021). This chapter has been substantially altered to improve clarity and should be referred to in lieu of the older published article if possible.

Emerson, R. W. (1836). *Nature*. James Munroe and Company.
Husserl, E. (1917/1964). In M. Heidegger (Ed.), *On the phenomenology of the consciousness of internal time* (1893–1917) (J.S. Churchill. Trans.). Indiana University Press.
Kent, L., & Wittmann, M. (2021). Time consciousness science and its theories. *Neuroscience of Consciousness, 7*(2), niab011. https://doi.org/10.1093/niab011
Lowe, E. J. (1987). Lewis on Perdurance versus endurance. *Analysis, 47*, 152–154.
McKenna, C. A. (2021). Don't go chasing waterfalls: Motion aftereffects and the dynamic snapshot theory of temporal experience. *Review of Philosophy and Psychology, 12*(4), 825–845.
McKenna, C. A. (2023). Agency and the successive structure of time-consciousness. *Erkenntnis, 88*(5), 2013–2034.
Merleau-Ponty, M. (1945). *Phenomenology of perception* (Colin Smith. Trans.). Routledge.
Seth, A. (2021). *Being you: A new science of consciousness*. Faber and Faber.

Chapter 2
The Phenomenology of Time-Consciousness

There are many things we should want from an account of time-consciousness. How well our theories can satisfy this list of demands helps to determine how good those theories are. The things we desire from our theories we can call *desiderata*. This chapter considers phenomenological desiderata, meaning those features of our subjective temporal experience that should be explained or explained away by our theories. Loosely put, these phenomenological features are aspects of what it is like for us to "live in time". If the theory does not explain actual phenomenological features, or, worse, predicts that different subjective experiences should occur than those that do, then the theory needs to be revised. It is therefore imperative that the explanatory targets be well characterized and clarified.

The aim of this chapter is to identify the temporal features of our experience that a successful theory or theories should in some way be able to account for. Further analysis might reveal that some of these features overlap or have an asymmetrical dependence on other features; for instance, simultaneity may be subordinate to the larger phenomenon of temporal unity. For now, we can consider the phenomenology of time-consciousness in a deliberately uncritical way, without presuming excessive theoretical baggage. In this preliminary consideration,[1] the most obvious temporal features might include the following:

a. Passage
b. Events
c. Simultaneity
d. Continuity
e. Endurance
f. Unity
g. Duration

[1] The order of this list is not meant to suggest anything about primacy, centrality, obviousness, etc., but is just one plausible order that these features might occur to someone pre-theoretically.

C. A. McKenna, *Constructing the Present*, Studies in Brain and Mind 24,
https://doi.org/10.1007/978-3-031-89571-5_2

h. Succession
i. Ubiquity
j. Variability

These features will each be described in detail in this chapter. While the above list[2] represents obvious phenomenological features of temporal experience, there are other more contentious aspects that philosophers and psychologists have proposed in the past. These have included features of the underlying structure of our experience that are perhaps less immediately apparent to us as subjects. An example is the so-called *specious present.* The doctrine of the specious present, popularized by William James, refers to the proposition that the experienced present is not instantaneous, as is often presumed, but rather comprises a finite *span* of time (James, 1890, pp. 609–610).

Another controversial proposal is Husserl's contention that the phenomenology of subjective time exhibits a *tripartite structure* including what he calls *retentions, primal impressions,* and *protentions* (Husserl, 1917). For Husserl, anticipations of the future and traces of the past are an essential feature of the way we experience events. Following Husserl, Merleau-Ponty proposed phenomenological *presence-in-absence*, referring to the notion that the unrealized future and nonexistent past are nonetheless present in experience in some way (Merleau-Ponty, 1945). The present work is committed to the veracity of the phenomenological features introduced by James, Husserl, and Merleau-Ponty and attempts, as most other contemporary views also do, to explain these features as well as the others. The reasons for taking such features seriously will be discussed in greater detail in later chapters, while this one will focus on less controversial features of temporal phenomenology.

a. **Passage**

Augustine was one of the first philosophers in the Western tradition known to have deeply reflected on time as a subjective phenomenon, and his analysis provides us with important insights that remain relevant for contemporary theories of temporal experience. Augustine recognized that, "It is in you, O mind of mine, that I measure the periods of time. Do not shout me down that it exists [objectively][editor's note retained]," (*Confessions*, Book XI, Chapter XXVII). Augustine thus makes a distinction between how time appears to us and the dynamics of the mind-independent world: "I measure as time present the impression that things make on you [the mind] as they pass by and what remains after they have passed by—I do not measure the things themselves which have passed by and left their impression on you," (ibid.).

[2] Various authors have emphasized or left out some of the features presented here. Ernst Pöppel, for instance, mentions only duration, non-simultaneity (implying also simultaneity), temporal order, past and present, and change, which, by his lights, subsumes the passage of time (Pöppel, 1978).

In other words, we feel time passing, whether or not it actually does. Augustine's observation of passage makes use of a common spatial metaphor,[3] with the experienced passage of time seeming to involve a kind of "motion" with a future-facing trajectory in which we are inexorably swept up. This feeling of passage has led to many of our best-known metaphors for time, from the "river" of Heraclitus to the "stream of consciousness"[4] and the oft-invoked folk imagery of paths, roads, ships, and journeys through life.[5]

Despite its apparent universality, the sentiment that we feel time passing is not wholly uncontroversial. Simon Prosser (2016) and Bradford Skow (2011), for instance, believe the experience of the passage of time is an illusion. Further, Prosser thinks the very idea of passage is incoherent. Regardless of whether the passage of time turns out to be an illusion or not, a good theory of subjective time must explain why it *seems to us* as if time passes. In other words, to what do we owe this deeply held and nearly universal intuition that experienced time has an essential movement-like quality and trajectory to it? While a natural answer to this question is just that it does in fact pass for us, there are other possibilities.

Even in antiquity, Augustine saw that the subjective passage of time presents serious philosophical difficulties. On this subject, he posed the following series of questions:

> [H]ow is it that there are the two times, past and future, when even the past is now no longer and the future is now not yet? But if the present were always present, and did not pass into past time, it obviously would not be time but eternity. If, then, time present—if it be time—comes into existence only because it passes into time past, how can we say that even this is, since the cause of its being is that it will cease to be? Thus, can we not truly say that time *is* only as it tends towards nonbeing? (*Confessions*, Book XI, Chapter XIV).

If the past is no longer and the future is not yet in existence, the question thus arises of how we can experience a *flow* of time. We have a feeling that, even though the past and the future are nonexistent, nonetheless they form a part of our experience in that the present is felt to constantly slip away into the past and the future is felt to constantly become the present. While the problem he identifies is philosophically significant, for the moment it is enough merely to highlight that Augustine has seemingly hit upon a phenomenological datum.

Beyond the holistic feeling of passage, there is a further specification of experiential "passage" involving the direct perception of change. C. D. Broad described a

---

[3] There is some evidence that in conceptualizing time humans have hijacked parts of the brain usually employed in the service of spatial mapping, so it's no surprise that the most natural metaphors we reach for when talking about time are spatial ones (see Gijssels & Casasanto, 2017).

[4] The phrase "stream of consciousness" may have been coined by Alexander Bain (1855, p. 359), but is widely attributed to William James (1890), who popularized the concept. The notion of the specious present, which comes from E. Robert Kelly (Anonymous, 1882), known to James as "E. R. Clay", followed a similar historical course.

[5] For example, F. Scott Fitzgerald ends *The Great Gatsby* with a particularly melancholic metaphor: "So we beat on, boats against the current, borne back ceaselessly into the past," (Fitzgerald, 1925/2004, p. 180).

now rather famous example of the hands of an analog clock. According to Broad, we do not merely perceive that the second hand was in one place and then, later, is in another, different place. Rather, there seems to be a phenomenology of change itself—it is a directly experienced feature of the second hand that it is in motion. As Broad puts it, "…we do not merely notice that something *has* moved or otherwise changed; we also often see something *moving* or *changing*," (Broad, 1923, p. 351). This stands in contrast to the hour hand, which is never *perceived* to be changing but only *believed* to have changed based on an inference from the difference between its spatial locations at different times. The experience of the second hand thus involves a distinctive phenomenology of change while the experience of the hour hand does not. Whether the same mechanism underpins the phenomenology of change for the objects of experience and the general phenomenology of time passing is a subject we will take up later, but suffice it to say for now that caution is warranted to avoid conflating these two experiences.

b. **Events**

One of the most obvious features of subjective experience—so obvious it is not often remarked upon—is that *things happen*. Experiences of things occur at certain times and over certain lengths of time. Furthermore, temporally extended processes, like exploding fireworks, seem to be grouped together such that we are aware of them as one process, although we can break a process into component processes as well (one firework explosion made up of smaller explosions for example). We can call these experiential processes that occur at times and over time "events." Straying from the phenomenology momentarily for the sake of elucidation, we might usefully think of events as Barbara Tversky and Jeffrey Zacks do: as subjectively experienced "chunks" of information, which, in perception, amount to an "integrated unit of space and time," (Tversky & Zacks, 2013, p. 83). There seem to be phenomenological boundaries to these processes. Vague and fuzzy though the boundaries might sometimes seem to us, nonetheless we apprehend a difference between the exploding firework and the subsequent stillness of the night sky.

c. **Simultaneity**

Experiences do not always occur serially—they can also feel as if they are occurring at the same time: for example, visual experiences of mouths moving and auditory experiences of the sounds they produce. At any given time, many experiences are simultaneous, from ongoing ambient noise to the tactile response of fingers on a keyboard, to the concurrent internal monologue and the background sensation of over-caffeinated inner restlessness. Consciousness is always marked by a jostling cacophony of experiences. While there will be more to say about simultaneity in **section f**, where it will return under the guise of "synchrony", simultaneity is a more introspectively obvious temporal feature than the more theoretically loaded "multi- and intra-sensory temporal integration" and so occupies a position closer to the beginning of this exposition.

d. **Continuity**

Not only do we feel as though we are moving through time into the future, but there is also a distinct feeling of continuity between what has just happened, what is happening, and what will happen. There is a first-personal subjective continuity (we are continuous with both our past and future selves) and continuity between all the other events that we experience through time. As William James put it many centuries after Augustine, subjective continuity means that, "even where there is a time-gap the consciousness after it feels as if it belonged together with the consciousness before it, as another part of the same self… [and] the changes from one moment to another in the quality of the consciousness are never absolutely abrupt," (James, 1892, p. 231). A full accounting of subjective time should ideally also explain how this feat—the problem of continuity—is pulled off.

Experiences happen over very short timescales, very long timescales, and timescales of varying length in between. What's more, experiences that occur over very short timescales (say, the time it takes to inadvertently burn oneself on a saucepan) seem phenomenologically distinct from experiences that occur over very long timescales (say, an overnight train journey). Nonetheless, we don't seem to experience any discontinuity between experiences at different timescales. On the contrary, our experiences feel *seamless* and it's hard to say when one ends and another begins. But how do we achieve this smooth integration of timescales—for example the continuity between timescales like that of the proposed specious present of 200 ms and a few seconds? Wanja Wiese has called this problem "The Interface Question" and formulates it this way: "How can I experience recent events as being seamlessly connected to present events?" (Wiese, 2017). Ideally, our theory of temporal experience should be able to provide an answer to this question.

e. **Endurance**[6]

In our experience, events and objects exhibit *endurance* or *persistence*, which is to say, we are aware of them lasting for some non-instantaneous length of time. A classic example of an enduring event, given by Sean Kelly, is an experience of a singer holding a note (Kelly, 2005, p. 228). In contrast, Wanja Wiese gives the example of a surprising "bright flash" as an example of a "punctual event [which] is not experienced as part of an enduring event," (Wiese, 2017, p. 16). It is debatable whether such events as a flash of light are really experienced without the event seeming to have any endurance whatsoever, but we may remain agnostic on this point for now.

Regardless of the possibility of punctual events, Wiese makes an important further point about the relationship between continuity and endurance that is more theoretical. Wiese contends that "continuity implies endurance: when we experience a temporal continuum, we experience a dynamic event, in which a higher-order event is experienced as enduring through change," (ibid.). The phenomenon of continuity is therefore dependent on the phenomenon of endurance—any continuity of

[6]This discussion concerns the apparent phenomenological feature of events that they seem like they *last* and not the metaphysical concepts of endurance and perdurance, which concern ways that identity can persist through time.

events would be impossible without their ability to endure in our experiences because events must endure to subsume other events that are felt to be continuous in virtue of their being subsumed by the enduring, higher-order event. There is thus a kind of "nesting" of enduring events that allows for continuity. Wiese's example of an event experienced in a smoothly temporally continuous way is the sound of a musical instrument playing notes "legato" as opposed to "staccato," (ibid.). The note events are perceived as an uninterrupted "*continua*" but also as belonging to a single, overarching higher-order event enduring through time, without which there would be no experience of continuity.

The experience of endurance is therefore quite central to temporal phenomenology. However, experiencing something *as* enduring is not simply a matter of events or objects persisting in the external world: we must somehow apprehend these things as being so. Persistent objects in the external world need not be perceived that way, after all; indeed, we can imagine experiencing the world in a strobe-light fashion fairly easily, and there are disorders of temporal awareness that involve such unpleasant occurrences.[7] A central phenomenon to be explained by a good theory of temporal experience is therefore how we should understand this endurance through time of events in our experience.

f. **Unity**

Another important temporal feature of experienced events is that we experience events in a unified way. This unity is both multimodal and intramodal, both synchronic and diachronic. As Shaun Gallagher and Dan Zahavi express it, "Experiences never occur in isolation. We are not faced with a mere aggregate of temporal atoms. The stream of consciousness is an ensemble of experiences unified both at any time, and over time...," (Gallagher & Zahavi, 2008, p. 72). For example, an event like the explosion of a firework incorporates visual, auditory, and if one is standing close enough to the explosion perhaps even olfactory, somatic, and thermoceptive information into a single experience felt as one event at a particular time and over a period of time. Despite significant differences in processing speed across the senses and differences in the speed of signals through the environment (photons move faster than airwaves, for example), the perceived event is still experienced as unified synchronically (i.e., at the same time) and diachronically (i.e., over time). Experiences also exhibit *intra*modal unity, i.e., a unity of the information perceived by one kind of sense perception, for example, the different colors of the firework, or the color and the motion of the explosion.

The phenomenon of unity naturally leads to the question of how we pull off such a feat, integrating all of this temporal information from an array of senses to make a whole, seamless experience. While multisensory integration in general faces a similar problem, there is a special problem when it comes to binding *temporal* information, rather than binding the contents of experiences. The latter involves binding content, like binding a voice to a face, while the former requires more. In temporal

[7] One such disorder is called akinetopsia and will be discussed in much greater detail in Chap. 5.

binding, not only is the content (voice and face) bound together, but also structural features, like the timing and duration of the content, are synced up. This suggests a different process is responsible for the timing of a voice and the movement of the face to be synchronized than in the traditional multisensory integration case. We can call the question of how such a process works the "temporal binding problem" (in contrast to the more general "binding problem").

To illustrate the complexity of the temporal binding problem, Jean Vroomen and Mirjam Keetels compare binding in the regular case to the temporal case. Binding is more likely to occur in general if "information from different modalities share[s] more (amodal) properties)," (Vroomen & Keetels, 2010, p. 871). If many properties are shared, then there is a high probability that the source of the information is the same across modalities. This principle is known as the "assumption of unity," (ibid.). Temporal properties like simultaneity are the most important shared properties in determining whether multisensory information should be bound together (ibid.). However, when it comes to binding temporal information itself to establish synchrony in perception, the assumption of unity breaks down on pain of circularity. The normal recourse to shared temporal properties becomes uninformative when temporal properties are the properties at issue in the case of temporal integration. There are further complications: besides the fact that, "no sense organ registers time on an absolute scale" the way a retina is sensitive to wavelengths of light, there are also "differences in physical and neural transmission times," for example the speeds of sound and light and the immediate transmission of information in the case of the sense of touch (ibid.).

To date, no single mechanism has been described that can explain how synchrony is established in experience. While explaining synchrony would be a desirable feature of a theory of subjective time, it may be that the process of integration is unrelated to the mechanism responsible for other aspects of temporal phenomenology like continuity and succession. For this reason, temporal unity is a phenomenological desideratum that requires special attention.

Note that the kind of unity described here is narrowly circumscribed, relating only to the binding of experiences together. In a wider sense, it is often commonly thought that all the various kinds of temporal phenomena, including "simultaneity, successiveness, temporal order, duration, and temporal perspective" are unified into a single experience of time (Block & Zakay, 2001, p. 59). However, as Barbara Sattler argues with reference to the Ancient Greek conception of time, such a sense of the unity of qualitatively different temporal phenomena has not always been so obvious. Sattler claims that intellectual history provides strong evidence that "our notion of time as unified is not something we gain directly from experience," but rather a malleable "interpretative or theoretical overlay" that has changed through the course of history (Sattler, 2017, p. 19). If this is correct, it serves as a reminder that we must be cautious about which temporal features we take as basic.

g. **Duration**

Another aspect of our experience of time is that we are conscious of intervals, which is to say durations, lengths, or spans of time. Beyond mere awareness of intervals, most of us can measure them as well. But this phenomenological fact introduces problems of its own. Augustine considers that we are wrong to say, for example, that a hundred years is a long time past, because what is past no longer exists, and what does not exist cannot be long. And yet it won't do to say it *was* long either, because that implies a century can happen, when it cannot, at least not all at once. determines that no time can be present, and thus, no time can exist, except an infinitesimally small point. As he explains, "If any fraction of time be conceived that cannot now be divided even into the most minute momentary point, this alone is what we may call time present. But this flies so rapidly from future to past that it cannot be extended by any delay. For if it is extended, it is then divided into past and future. *But the present has no extension whatever* [emphasis added]," (*Confessions*, Book XI, Chapter XV). "And yet," Augustine writes, "we do perceive intervals of time, we do compare them with each other, and we say that some are longer and others are shorter," (*Confessions*, Book XI, Chapter XVI).

Thus, a conundrum arises: on the one hand we seem to have the brute phenomenological fact that we are conscious of intervals of time, and on the other we have strong *a priori* grounds for holding that the present cannot have extension, because if it did it wouldn't be present. How is it possible that intervals form a part of our experience? Augustine's frustration with this antinomy is apparent in his colorful prose: "My soul burns ardently to understand this most intricate enigma," (ibid.). As demonstrated by debates between "snapshot" and "specious present" theorists, discussed in Chaps. 4 and 5, Augustine's burning worry has continued to vex philosophers down to the present day. I will argue in these later chapters, however, that the interval conception of the present is on the right track and the duration of this interval we call the present itself calls out for some further explanation.

h. **Succession**

An apparently obvious fact about our experience of time that nevertheless requires some explanation is that we experience events one after another, in succession. As I understand it, succession does not necessitate a train of *discrete* (i.e., atom-like, staccato, disjoint) experiential events, though it may be simplest to speak in this way about complex phenomena. Succession may also describe the state of affairs obtaining in a dynamic continuous process unfolding over time.[8] At first glance it might seem there's nothing extraordinary about succession—the universe appears to exhibit a successive quality itself, at least if common sense is believed, where the way things are at one time is followed by a way things are at another time, which may or may not be the same in all respects excepting the transposition along the temporal dimension.

The problem is that, while succession may be a feature of the universe we inhabit, a mere series of events is neither necessary nor sufficient to generate a

---

[8] I note this here also to make plain that whatever I suggest later in the book about succession is certainly not intended to endorse a discrete model of experience (or, for that matter, a similar view about reality), although I may talk in terms of discrete events for the sake of convenience. In fact, I find the process view to be a much more natural description of subjective experience.

phenomenology of succession. As William James put it, "a succession of feelings, in and of itself, is not a feeling of succession," (James, 1890, p. 591). Dan Dennett illustrates the point further by disentangling representational content from the representational vehicle of that content: "the representing by the brain of *A before B* does not have to be accomplished by: first: a representing of A, followed by: a representing of B. The phrase 'B after A' is an example of a (spoken) vehicle that represents A as being before B, and the brain can avail itself of the same freedom of temporal placement," (Dennett, 1993, p. 149). For this reason, it is within the realm of possibility that the actual temporal order or metaphysical structure of the universe around us is completely different (even reversed!) from what we experience. We needn't even commit ourselves to representational ways of thinking for James' problem to stick—we need only consider that events that occur in succession too quickly (like a bullet) or too slowly (like an hour hand) do not result in a feeling of succession (Kiverstein, 2010, p. 161). No matter the temporal structure of the universe or the status of representations, the feeling of succession stands in need of some explanation.

i. **Ubiquity**

Temporal phenomenology is not unique to sensory perception. Our imaginings, our dreams, our recollections of memories—in short, all conscious experiences—are temporally structured. It is quite difficult to imagine how this could not be the case, but nonetheless something must be occurring physiologically to make it so. The global presence of temporal phenomenology throughout consciousness is not only something to be explained, but it is also an indicator of the kind of mechanism we should be looking for—namely, a mechanism that is not isolated to a specific region or function but likely low-level and widely distributed.

j. **Rates and Variability**

Another salient phenomenological feature of temporal experience is that time seems to pass at a certain rate, and this rate can feel highly variable. The idea that these rates or speeds at which things happen fluctuate frequently is enshrined in common sense, for example in the old adage "time flies when you're having fun." As it turns out, copious experimental attention to subjective temporal variability has revealed that not only does time fly when we're having fun, it is also affected by innumerable other variables. For empirical investigators, timing tasks have been the most fruitful and most studied experimental paradigms in the field of subjective time research. This is because in these tasks we can compare the subject's behavior or reports to an objective measure (clock time).

The modulation of rates of passage is often subjectively discernable (and renowned in folk wisdom) while at other times the changes "go unnoticed, but nevertheless show relatively strong effects when measured," (Noreika et al., 2014, p. 536). Factors affecting subjective rates of passage apparently include but are not limited to age (Wittmann, 2005), gender (Espinosa-Fernández et al., 2003), body temperature, heart rate, time of day (Hancock & Hancock, 2014), interval duration (Hancock & Rausch, 2010), awe (i.e., "the emotion that arises when one encounters something so strikingly vast that it provokes a need to update one's mental schemas (Keltner & Haidt, 2003)" (Rudd et al., 2012)), cultural differences (Arman & Adair, 2012), interference from other modalities (Chen & Yeh, 2009), various emotions (Droit-Volet & Meck, 2007), personality traits like impulsivity (Wittmann & Paulus, 2008), the amount of change or variation that occurs within an interval (Hansen & Trope, 2013), migraine headaches (Vicario et al., 2014), exogenous oxytocin (Colonnello et al., 2016), bare physiological arousal (Mella et al., 2011), repetition (Matthews & Gheorghiu, 2016) and routine (Avni-Babad & Ritov, 2003), subjectively felt speed of movement (Zhang et al., 2014), hypnosis (Bowers, 1979), and of course boredom (Noreika et al., 2014). Additionally, drugs of many kinds are known to influence the experience of time, including marijuana, alcohol, stimulants (Meck, 1996), depressants (Shahabifar & Movahedinia, 2016), and psychedelics (Aronson, 1959). Depending on the variable in question, the rate or speed of the passage of time is felt to either slow down (time dilation) or speed up (time contraction)—for example, having fun results in time contraction (time flies) while a boring statistics lecture would result in time dilation (slowing to a crawl).

What can we take away from this extreme variability? For one thing, temporal variability is the rule rather than the exception. Subjective rates are in a constant state of flux depending on a huge array of factors. Because of this, we might consider that the most plausible mechanism would occur at a very low level in the implementation to be modulated by such a wide variety of different variables. Furthermore, the mechanism or mechanisms involved could be widely distributed throughout the brain and/or entire central nervous system to be sensitive to said variables, perhaps extending well beyond the systems responsible for conscious experience.

Studies on the variability of subjective time must be handled cautiously. Researchers must always be on guard against inadvertently conflating judgment, memory, and the actual experience, which are not necessarily always in alignment. Unfortunately, these processes are difficult to disentangle experimentally, especially when a subject is asked to self-report. In such cases it can be impossible to determine definitively if the effect is on a subject's memory of the event, the original experience of the event, or both. Other issues concern whether the effects are task-related or would remain in natural conditions and also whether the effects vary or even reverse depending on the timescales considered.

So far, we have been describing temporal phenomenology under more or less normal conditions. However, there are well-documented cases where the experience of time is known to deviate substantially from the neurotypical case. These deviations represent explananda for a theory of temporal experience as well and may also offer clues as to what is going on at the processing level.

We can divide phenomenological deviations into several categories. First, there are the chronic mental disorders, which, although often associated with other disturbances to normal functioning, have implications for subjective time. A non-exclusive list of disorders that purportedly affect temporal phenomenology would include schizophrenia (Carroll et al., 2008), depression (Bech, 1975), bipolar disorder (Bschor et al., 2004), ADHD (Barkley et al., 2001), anxiety disorders (Mioni et al., 2016), personality disorders (Berlin et al., 2005), autism (Allman et al., 2011), and dyslexia (Ben-Artzi et al., 2005). Second, there are disorders specifically related to the experience of time, like dyschronometria and akinetopsia. These are uncommon and usually occur as symptoms of a larger problem or a direct result of lesions to specific areas of the brain. Third are various supposed "illusions of time" which occur in normal subjects as well as other temporary alterations to the normal experience of time instigated by external factors like hypnosis or psychoactive drugs. We will now turn to investigating some of these deviations in greater detail.

2.1 General Disorders Affecting Temporal Experience

Many disorders of the mind seem to affect subjective time. There are however methodological difficulties common to all of them that make any assessment of the impact of these disorders challenging. Because it is not possible to experience what the subject is experiencing, we have to rely on inference to determine what's going in their perception.

As noted earlier, the disentanglement of experience, judgment, and memory is vital for empirical research but often impossible in common experimental paradigms. Self-reporting by the subject might seem like a good basis for inference, but this method invites complications. Rather than a guide to the subject's experiences at the time in question, self-reports might instead only indicate the subject's *memory* of the experience or ability to *imagine* future experiences. Indeed, if self-reporting were a good guide to subjective experience, we should expect the reports to match objective data collected from experimental timing tasks.

In fact, mismatches between the data collected in the experimental tasks and self-reporting are common in certain populations, a fact which was noted as far back as 1932 when Aubrey Lewis found accurate judgments on timing tasks often belied self-reported distortions of temporal experience (Lewis, 1932). A further worry is that, even in the experimental task, it is really the subject's ability to make judgments about their experience of time that is tested rather than directly testing their experience of time, and their experience and their ability judge it accurately might deviate. However, so long as other minds remain inaccessible to observers, we have to make do with suggestive but ultimately inconclusive inferences based on experimental timing tasks, self-reporting, and in some cases, neuroimaging and trans-cranial magnetic stimulation (TMS). If retrospective timing tasks, prospective timing tasks, and self-reporting coincide, we may consider this a suggestive but not definitive indicator that the data do actually reflect the subject's temporal experience.

2.1.1 *Schizophrenia*

Schizophrenics have problems estimating time intervals when compared to the general population. These problems can include "a breakdown of temporal order of external events or an altered speed of subjective versus objective time passage," (Noreika et al., 2014). Self-reporting has shown some variance in the distortion experienced by schizophrenics, from time dilation to rare reports of time stoppage (Freedman, 1974). It is hard to make sense of anecdotes of time stoppage. However, consistent overestimation of time intervals at all timescales in experimental tasks suggests time dilation is the most common subjective distortion experienced by schizophrenics regardless of subtype (e.g., paranoid vs. non-paranoid, chronic vs. non-chronic) (Tysk, 1983). Furthermore, fMRI studies have shown that schizophrenics exhibit "timing-specific fronto-thalamo-striatal dysfunction" when administered duration discrimination tasks (tasks where subjects judge differences between two durations) (Volz et al., 2001), which indicates a neurological underpinning to this population's altered sense of timing.

2.1.2 *Depression and Mania*

Self-reporting from those afflicted by depression seems to support the notion that they experience time dilation as well (Bschor et al., 2004). On the other hand, in mania, time contraction is reported. However, here there is a mismatch with objective testing: in one study, both manic and depressed experimental groups made judgments about the length of long intervals as if they were experiencing time dilation, with manic patients exhibiting greater overestimation of intervals than either the depressed or control groups (ibid.). Although sadness is also correlated with the overestimation of durations, the effects produced by sadness alone are much less pronounced than those observed in depressed or manic patients, indicating a further process is responsible for temporal modulation in those cases (Gil & Droit-Volet, 2009). In both schizophrenia and depression then, as well as other disorders not detailed here, there seem to be deviations in the normal functioning of the mechanism responsible for subjective time which result in distorted phenomenology and judgments of time compared to the normal case. A better understanding of how these disorders affect the brain could therefore also help to reveal the mechanism underlying temporal phenomenology.

2.2 Specific Disorders of Temporal Experience

2.2.1 Akinetopsia

Akinetopsia refers to a disorder that limits the ability to perceive motion, either entirely or to some degree. Akinetopsia can leave the affected person with experiences they describe as being like a series of static images or frames passing, similar to a very low frame-rate film or strobe light effect (stroboscopic vision). Although most cases are stroboscopic, in very rare circumstances those affected report certain moving things, like flowing liquid, to be static when they are not.

The neurophysiological situation in such cases is instructive. In one patient exhibiting the more extreme form of the disorder, researchers observed "bilateral lesions affecting the lateral temporo-occipital cortex and the underlying white matter" and concluded on this basis that, "movement vision is a separate visual function depending on neuronal mechanisms beyond the primary visual cortex," (Zihl et al., 1983). In particular, damage to region V5 (or MT) in the occipital lobe appears to be implicated in akinetopsia. In monkeys, cells in the V5 region "preferentially respond to the direction and speed of stimulus motion," (Heywood & Kentridge, 2009), while in humans, a region called hMT+, "adapts to motion of complex patterns while lower visual areas adapt to motion of the components (Huk & Heeger, 2002)," (Heywood & Kentridge, 2009).

The upshot of cases of akinetopsia is that they offer examples of temporal phenomenology and visual motion experience coming apart. Despite the inability to perceive motion and change, motion blind subjects continue to experience time passing along with all of the other phenomenal features of subjective time delineated in this chapter. In other words, while it may be that the flowing liquid appears not to move, it is not experienced as remaining static through time, whatever that would mean. This situation has a number of implications, which will be discussed in much greater detail in Chap. 5.

2.2.2 Dyschronometria

Dyschronometria refers to a sustained inability of a subject to accurately estimate durations of passing time. This phenomenon often presages the onset of dementia and can be caused by cerebellar ataxia. It is unclear exactly what the impaired process behind this disorder is, but it could be that, rather than a disorder of subjective time *per se*, it is instead a disorder of memory that affects the subject's ability to judge and report on passing intervals. That is to say, it could be that in many cases of dyschronometria the subject forgets they were supposed to be engaged in measuring a duration before they are able to measure it. This would mean that the explanation for dyschronometria should, properly speaking, come from a theory of memory and not necessarily from a theory of time-consciousness *per se*.

2.3 Temporary Distortions

2.3.1 Psychoactive Drugs

Psychoactive substances are often associated with extreme distortions of time. Early studies on LSD use, conducted in laboratory settings, produced a range of striking subjective reports, for example: "there was a feeling of eternity, of timelessness"; "both past and future seemed very far away"; "[it was] as though there were not time at all, and yet the time seemed to pass slowly," (Kenna & Sedman, 1964). In objective timing tasks those that took LSD consistently underestimated the duration of intervals across a range of timescales, indicating time contraction, despite subjective reports of time feeling slowed down (Aronson, 1959). Another interesting phenomenon from LSD studies was that, in one study (Boardman, 1957), subjects who took LSD and were asked to estimate whether a sound had lasted a second or not reported that, "it was difficult to maintain a stable concept of a second under LSD," (Noreika et al., 2014, p. 536). Unfortunately, many early studies suffered from poor methodology and few participants, making the results difficult to interpret. Compounding these problems is that replication is now virtually impossible following the moratorium on LSD studies. Suffice it to say that LSD and other psychoactive drugs are capable of profoundly altering temporal experience.[9]

2.3.2 Illusions

There are several kinds of perceptual illusions that might have implications for a theory of temporal experience. Among these, the most notable and bewildering are probably the so-called "postdictive" effects. These are counterintuitive cases where events occurring after a stimulus seemingly affect the subject's perception of that earlier stimulus. Some of the most prominent postdictive effects are the illusions of apparent motion (also known as the phi phenomenon) (Kolers, 1972, pp. 1–15), metacontrast marking (Bachmann & Francis, 2014), the flash-lag effect(Mackay, 1958; Eagleman and Sejnowski, 2000), and the cutaneous rabbit illusion (Geldard & Sherrick, 1972). In the case of apparent motion, two stimuli, separated in space, are presented in succession, and the subject perceives that the earlier stimulus is moving to the place of the other. This phenomenon is extremely robust and allows us to perceive successive static frames as continuous movement, like in a flipbook or while watching a film or animation. Mundane though such experiences seem to us, they conceal a paradox, namely that our perception of motion seems to depend on the existence of a future stimulus.

---

[9] See Wittmann (2018) for more on altered states of time-consciousness, including research on time dilation and contraction related to a variety of major drug classes.

A particular version of apparent motion, known as color-phi, has garnered special attention because of Dan Dennett's use of the phenomenon in arguing against the doctrine of the Cartesian Theater, which is the idea there is a perceptual endpoint in our mind that we watch like an audience watches a performance (Dennett, 1993, p. 114). Even more so than run-of-the-mill apparent motion, color phi is a striking demonstration of "postdiction". In color phi, two spatially separated stimuli (in practice, dots) of different colors are presented, one after the other. Subjects not only report motion of the stimulus but a change in color that takes place midway through the motion of the stimulus. Of course, neither motion nor a color change halfway through actually occurred. Yet it seems as though the color and placement of the final stimulus influences the perception of what is happening before its onset (the motion and color change of the first dot), which is a highly counterintuitive result. For the moment, we needn't discuss all permutations of postdictive effects. The important thing is that they all share the feature of seeming to show that later events can have an impact on "present" perceptions. Our theory of temporal experience must therefore accommodate these bizarre phenomena.

Although postdictive effects are sometimes considered illusions of time, we should be clear that this may not be entirely accurate nomenclature. In illusions, there is a mismatch between perception and reality, such that we take our perception to be inaccurate or non-veridical—a misperception. Usually this involves perceiving the wrong properties of perceptual objects, for example, seeing solid pavement as liquid on a hot day. In the case of postdictive effects, however, as Valtteri Arstila points out, "the effects do not demonstrate that temporal properties per se (e.g., simultaneity and duration of stimuli) are misperceived. Instead, the temporally interesting feature of these cases depends on our view concerning the 'normal' order of neural processing and the timing of experiences," (Arstila, 2015, p. 206). In other words, the temporal features are preserved—the presented stimuli occur in succession, over a certain amount of time, etc.—the strangeness is just that we don't expect earlier perceptions to be affected by later events. The surprise is therefore a function of the way we expect processing to work (i.e., sequentially with a definite perceptual endpoint), rather than a misperception of any temporal feature of the object. Of course, it must be said that in a non-temporal sense, these are still illusions, but illusions of the senses rather than illusions of time. In the apparent motion case, for instance, we have visual misperceptions, seeing an object as changing colors, occupying various spatial locations, and moving, even though the object itself does not exhibit any of those behaviors.

Still, insofar as it appears later events are influencing earlier events in a counterintuitive fashion, it seems postdictive effects have at least something to do with subjective time or the temporal processing of events. Furthermore, in some of these illusions, like the flash-lag (of which more in later chapters) it can seem to be the case that an event takes place at a different time than it really did. If there is really an illusion of event order going on such that the actual event order is experienced non-veridically, the illusion could plausibly be understood as an illusion of subjective time.

2.4 Conclusion

This chapter has described various phenomenological features that a good theory of temporal experience should seek to explain. These include the feeling of time passing, experiences of events, simultaneity, continuity, endurance, intervals, temporal unity, succession, ubiquity throughout subjective experience, variable rates and timescales, and occasionally abnormal temporal phenomenology. We find this temporal phenomenology not only pertaining to our immediate perception of the world around us, but also in our experiences of memory, imagination, and all other areas of conscious experience. Contemporary theories of temporal experience can be evaluated against their ability to provide explanations for the phenomena identified here. Before moving on to such an evaluation however, further constraints on the space of theoretical possibilities for such theories should be considered, and this will be the subject of the next chapter.

References

Allman, M. J., DeLeon, I. G., & Wearden, J. H. (2011). Psychophysical assessment of timing in individuals with autism. *American Journal on Intellectual and Developmental Disabilities, 116*, 165–178. https://doi.org/10.1352/1944-7558-116.2.165

Anonymous (E. Robert Kelly). (1882). *The alternative: A study in psychology*. Macmillan and Co.

Arman, G., & Adair, C. K. (2012). Cross-cultural differences in perception of time: Implications for multinational teams. *European Journal of Work and Organizational Psychology, 21*, 657–680. https://doi.org/10.1080/1359432X.2012.662488

Aronson, H. (1959). Influence of lysergic acid diethylamide (LSD-25) on subjective time. *Archives of General Psychiatry, 1*, 469. https://doi.org/10.1001/archpsyc.1959.03590050037003

Arstila, V. (2015). Keeping postdiction simple. *Consciousness and Cognition*, 38, 205–216. https://doi.org/10.1016/j.concog.2015.10.001

Avni-Babad, D., & Ritov, I. (2003). Routine and the perception of time. *Journal of Experimental Psychology: General, 132*, 543–550. https://doi.org/10.1037/0096-3445.132.4.543

Bachmann, T., & Francis, G. (2014). *Visual masking: Studying perception, attention, and consciousness*. Academic.

Bain, A. (1855). *The senses and the intellect*. J. W. Parker and Son.

Barkley, R. A., Murphy, K. R., & Bush, T.. (2001). Time perception and reproduction in young adults with attention deficit hyperactivity disorder. *Neuropsychology, 15*, 351–360. https://doi.org/10.1037//0894-4105.15.3.351

Bech, P. (1975). Depression: Influence on time estimation and time experience. *Acta Psychiatrica Scandinavica, 51*, 42–50. https://doi.org/10.1111/j.1600-0447.1975.tb00211.x

Ben-Artzi, E., Fostick, L., & Babkoff, H. (2005). Deficits in temporal-order judgments in dyslexia: Evidence from Diotic stimuli differing spectrally and from dichotic stimuli differing only by perceived location. *Neuropsychologia, 43*, 714–723. https://doi.org/10.1016/j.neuropsychologia.2004.08.004

Berlin, H. A., Rolls, E. T., & Iversen, S. D. (2005). Borderline personality disorder, impulsivity, and the orbitofrontal cortex. *American Journal of Psychiatry, 162*, 2360–2373. https://doi.org/10.1176/appi.ajp.162.12.2360

Block, R., & Zakay, D. (2001). Retrospective and prospective timing: Memory, attention, and consciousness. In C. Hoerl & T. McCormack (Eds.), *Time and memory: Issues in philosophy and psychology* (pp. 59–76). Oxford University Press.

Boardman, W. K. (1957). Effects of lysergic acid diethylamide (LSD) on the time sense of normals: A preliminary report. *A.M.A Archives of Neurology & Psychiatry, 78*, 321. https://doi.org/10.1001/archneurpsyc.1957.02330390103013

Bowers, K. S. (1979). Time distortion and hypnotic ability: Underestimating the duration of hypnosis. *Journal of Abnormal Psychology, 88*, 435–439. https://doi.org/10.1037/0021-843X.88.4.435

Broad, C. D. (1923). *Scientific thought: A philosophical analysis of some of its fundamental concepts*. Kegan Paul.

Bschor, T., Ising, M., Bauer, M., Lewitzka, U., Skerstupeit, M., Muller-Oerlinghausen, B., & Baethge, C. (2004). Time experience and time judgment in major depression, mania and healthy subjects: A controlled study of 93 subjects. *Acta Psychiatrica Scandinavica, 109*, 222–229. https://doi.org/10.1046/j.0001-690X.2003.00244.x

Carroll, C. A., Boggs, J., O'Donnell, B. F., Shekhar, A., & Hetrick, W. P. (2008). Temporal processing dysfunction in schizophrenia. *Brain and Cognition, 67*, 150–161. https://doi.org/10.1016/j.bandc.2007.12.005

Chen, K.-M., & Yeh, S.-L. (2009). Asymmetric cross-modal effects in time perception. *Acta Psychologica, 130*, 225–234. https://doi.org/10.1016/j.actpsy.2008.12.008

Colonnello, V., Domes, G., & Heinrichs, M. (2016). As time goes by: Oxytocin influences the subjective perception of time in a social context. *Psychoneuroendocrinology, 68*, 69–73. https://doi.org/10.1016/j.psyneuen.2016.02.015

Dennett, D. (1993). *Consciousness explained*. Penguin.

Droit-Volet, S., & Meck, W. H. (2007). How emotions colour our perception of time. *Trends in Cognitive Sciences, 11*, 504–513. https://doi.org/10.1016/j.tics.2007.09.008

Eagleman, D. M., & Sejnowski, T. J. (2000). Motion integration and postdiction in visual awareness. *Science*, 287, 2036–2038.

Espinosa-Fernández, L., Miró, E., Cano, M., & Buela-Casal, G. (2003). Age-related changes and gender differences in time estimation. *Acta Psychologica, 112*, 221–232. https://doi.org/10.1016/S0001-6918(02)00093-8

Fitzgerald, F. S. (1925/2004). *The great Gatsby*. Scribner.

Freedman, B. J. (1974). The subjective experience of perceptual and cognitive disturbances in schizophrenia: A review of autobiographical accounts. *Archives of General Psychiatry, 30*, 333. https://doi.org/10.1001/archpsyc.1974.01760090047008

Gallagher, S., & Zahavi, D. (2008). *The phenomenological mind*. Routledge.

Geldard, F. A., & Sherrick, C. E. (1972). The cutaneous "Rabbit": A perceptual illusion. *Science, 178*(4057), 178–179.

Gijssels, T., & Casasanto, D. (2017). Conceptualizing time in terms of space: Experimental evidence. In B. Dancygier (Ed.), *The Cambridge handbook of cognitive linguistics* (pp. 651–668). Cambridge University Press. https://doi.org/10.1017/9781316339732.041

Gil, S., & Droit-Volet, S. (2009). Time perception, depression and sadness. *Behavioural Processes, 80*, 169–176. https://doi.org/10.1016/j.beproc.2008.11.012

Hancock, P., & Hancock, G. (2014). The effects of age, sex, body temperature, heart rate, and time of day on the perception of time in life. *Time & Society, 23*, 195–211. https://doi.org/10.1177/0961463X13479187

Hancock, P. A., & Rausch, R. (2010). The effects of sex, age, and interval duration on the perception of time. *Acta Psychologica, 133*, 170–179. https://doi.org/10.1016/j.actpsy.2009.11.005

Hansen, J., & Trope, Y. (2013). When time flies: How abstract and concrete mental construal affect the perception of time. *Journal of Experimental Psychology: General, 142*, 336–347. https://doi.org/10.1037/a0029283

Heywood, C., & Kentridge, R. W. (2009). Akinetopsia. In T. Bayne, A. Cleeremans, & P. Wilken (Eds.), *The Oxford companion to consciousness*. Oxford University Press.

Huk, A. C., & Heeger, D. J. (2002). Pattern-motion responses in human visual cortex. *Nature Neuroscience, 5*, 72–75. https://doi.org/10.1038/nn774

Husserl, E. (1917/1964). In M. Heidegger (Ed.), *On the phenomenology of the consciousness of internal time* (1893–1917) (J.S. Churchill. Trans.). Indiana University Press.

James, W. (1890/1980). *The principles of psychology*. Henry Holt.

James, W. (1892). *Psychology*. Fawcett Premier.

Kelly, S. (2005). Temporal awareness. In D. W. Smith & A. L. Thomasson (Eds.), *Phenomenology and philosophy of mind* (pp. 222–234). Oxford University Press.

Keltner, D., & Haidt, J. (2003). Approaching awe, a moral, spiritual, and aesthetic emotion. *Cognition and Emotion, 17*, 297–314. https://doi.org/10.1080/02699930302297

Kenna, J. C., & Sedman, G. (1964). The subjective experience of time during lysergic acid diethylamide (LSD-25) intoxication. *Psychopharmacologia, 5*, 280–288. https://doi.org/10.1007/BF02341260

Kiverstein, J. (2010). Making sense of phenomenal unity: An intentionalist account of temporal experience. *Royal Institute of Philosophy Supplement, 67*, 155–181. https://doi.org/10.1017/S1358246110000081

Kolers, P. A. (1972). *Aspects of motion perception*. Pergamon Press.

Lewis, A. (1932). The experience of time in mental disorder. *Proceedings of the Royal Society of Medicine, 25*, 611–620.

MacKay, D. M. (1958). Perceptual stability of a stroboscopically lit visual field containing self-luminous objects. *Nature, 181*, 507–508.

Matthews, W. J., & Gheorghiu, A. I. (2016). Repetition, expectation, and the perception of time. *Current Opinion in Behavioral Sciences, 8*, 110–116. https://doi.org/10.1016/j.cobeha.2016.02.019

Meck, W. H. (1996). Neuropharmacology of timing and time perception. *Cognitive Brain Research, 3*, 227–242. https://doi.org/10.1016/0926-6410(96)00009-2

Mella, N., Conty, L., & Pouthas, V. (2011). The role of physiological arousal in time perception: Psychophysiological evidence from an emotion regulation paradigm. *Brain and Cognition, 75*, 182–187. https://doi.org/10.1016/j.bandc.2010.11.012

Merleau-Ponty, M. (1945). *Phenomenology of perception* (Colin Smith. Trans.). Routledge.

Mioni, G., Stablum, F., Prunetti, E., & Grondin, S. (2016). Time perception in anxious and depressed patients: A comparison between time reproduction and time production tasks. *Journal of Affective Disorders, 196*, 154–163. https://doi.org/10.1016/j.jad.2016.02.047

Noreika, V., Falter, C. M., & Wagner, T. M. (2014). Variability of duration perception: From natural and induced alterations to psychiatric disorders. In V. Arstila & D. Lloyd (Eds.), *Subjective time* (pp. 529–556). MIT Press.

Pöppel, E. (1978). Time perception. In R. Held (Ed.), *Handbook of sensory physiology*. Springer.

Prosser, S. (2016). *Experiencing time*. Oxford University Press.

Rudd, M., Vohs, K. D., & Aaker, J. (2012). Awe expands people's perception of time, alters decision making, and enhances well-being. *Psychological Science, 23*, 1130–1136. https://doi.org/10.1177/0956797612438731

Sattler, B. (2017). How natural is a unified notion of time: Temporal experience in early Greek thought. In I. Phillips (Ed.), *The Routledge handbook of philosophy of temporal experience* (pp. 19–29). Routledge.

Shahabifar, A., & Movahedinia, A. (2016). Comparing time perception among morphine-derived drugs addicts and controls. *Addiction and Health, 8*, 32–40.

Skow, B. (2011). Experience and the passage of time. *Philosophical Perspectives: Metaphysics, 25*, 359–387.

Tversky, B., & Zacks, J. M. (2013). Event perception. In D. Reisberg (Ed.), *The Oxford handbook of cognitive psychology* (pp. 83–96). Oxford University Press.

Tysk, L. (1983). Estimation of time and the subclassification of schizophrenic disorders. *Perceptual and Motor Skills, 57*, 911–918. https://doi.org/10.2466/pms.1983.57.3.911

Vicario, C. M., Gulisano, M., Martino, D., & Rizzo, R. (2014). The perception of time in childhood migraine. *Cephalalgia, 34*, 548–553. https://doi.org/10.1177/0333102413517774

Volz, H.-P., Nenadic, I., Gaser, C., Rammsayer, T., Häger, F., & Sauer, H. (2001). Time estimation in schizophrenia: An fMRI study at adjusted levels of difficulty. *Neuroreport, 12*, 313–316. https://doi.org/10.1097/00001756-200102120-00026

Vroomen, J., & Keetels, M. (2010). Perception of intersensory synchrony: A tutorial review. *Attention, Perception, & Psychophysics, 72*, 871–884. https://doi.org/10.3758/APP.72.4.871

Wiese, W. (2017). Predictive processing and the phenomenology of time consciousness: A hierarchical extension of Rick Grush's trajectory estimation model. In T. Metzinger & W. Wiese (Eds.), *Philosophy and predictive processing* (Vol. 26). MIND Group. https://doi.org/10.15502/9783958573277

Wittmann, M. (2005). Age effects in perception of time. *Psychological Reports, 97*, 921. https://doi.org/10.2466/PR0.97.7.921-935

Wittmann, M. (2018). *Altered states of consciousness: Experiences out of time and self* (P. Hurd, Trans.). MIT Press.

Wittmann, M., & Paulus, M. P. (2008). Decision making, impulsivity and time perception. *Trends in Cognitive Sciences, 12*, 7–12. https://doi.org/10.1016/j.tics.2007.10.004

Zhang, Z., Jia, L., & Ren, W. (2014). Time changes with feeling of speed: An embodied perspective. *Frontiers in Neurorobotics, 8*. https://doi.org/10.3389/fnbot.2014.00014

Zihl, J., von Cramon, D., & Mai, N. (1983). Selective disturbance of movement vision after bilateral brain damage. *Brain, 106*(2), 313–340.

Chapter 3
The Necessity of Experiential Succession

While the last chapter considered the pre-theoretical characteristics of temporal experience, this chapter will consider potential constraints on the phenomenal character of temporal experience. Specifically, this chapter will argue that experiential succession, mentioned previously as one of many subjective temporal phenomena, is in reality a necessary condition for agentive subjects in this world. As a result, we should look to place succession, indispensable as it is to experience, at the core of any theory we might produce to account for experience *at all*, let alone so-called "temporal experience," which, seen in this light, no longer appears as a subspecies of experience but rather provides the grounds for our existence as subjects.

In an oft-quoted passage, William James claims "the unit of composition of our perception of time is a *duration*, with a bow and a stern, as it were—a rearward- and a forward-looking end. It is only as parts of this *duration-block* that the relation of *succession* of one end to the other is perceived," (James, 1890, pp. 609–610). This quote and the section of *The Principles of Psychology* from which it is extracted have come to be seen as embodying an initial statement of the doctrine of the specious present, which is the idea that the experienced present has some kind of temporal extension. However, there is a further contention in the above quote that Jan Almäng (2014, p. 364) has noted. Besides the claim that the present is a "duration-block," as opposed to a "knife-edge," James also claims that our temporally extended present has a *successive structure*.

This latter claim about successive structure is often overlooked because we ordinarily take it for granted that experience is successive, whereas we ordinarily do not take it for granted that the perceived present has a duration. For Almäng, James' succession claim, when made precise, amounts to the following: "That with which we are perceptually presented is given as having temporal parts, which are presented as succeeding and preceding each other in time." (Almäng, 2014, p. 364). This characterization of the succession claim is a minimal description of how we contingently experience the world over time, asserting only what is meant to be obviously manifest.

C. A. McKenna, *Constructing the Present*, Studies in Brain and Mind 24,
https://doi.org/10.1007/978-3-031-89571-5_3

I think the succession claim can and should be stronger still. I will argue that, for agentive subjects at least, the successive structure of experience is not merely contingent but nomologically necessitated. I argue for constraining the possibility space of temporal experiences, endorsing what I call the Succession Requirement (SR), which holds the experiences of agentive subjects in our world must have the same basic temporal structure, namely, a successive structure. To be clear, this is not to say there is no variation in the content of temporal experience (subjective time dilation and contraction, for example, are well-documented[1]). Rather, SR holds that the basic successive *structure* of temporal experience must remain constant across subjects in this world that we would consider to be agents, where by "successive structure" I mean a temporal organization or *ordering* of the content of experience.

While the proposition that we experience the world in successive fashion may be very intuitive, the stronger modal claim that we *must* experience the world in this way is not. Just as it is not obvious that experienced succession is necessary for agency, non-agency also does not follow in an immediately obvious way from non-succession. Nonetheless, I maintain there are very compelling reasons to believe SR holds. I will focus on three of these here. The first is that successive experience is the only perspective from which an agent can navigate a world that includes causality[2] as a feature, as non-successive experiencers would not be able to stand in the appropriate causal relations to their actions. Second, a physically grounded informational asymmetry between past and future means that any non-successive experiencer would lack the requisite epistemic status for agency as we understand it. Third, only successive experiencers are capable of the kind of mental attitudes that ground agency—a fact rooted in the processive nature of experience.

The Succession Requirement, if it holds, has a number of surprising and significant implications. Perhaps the most surprising and least intuitive is that it implies a temporal "God's-eye-view" is actually sub-optimal in several ways. Agents like us occupy a privileged position in our temporal perspective, which is specifically adapted to handle the real physical asymmetries of the universe we inhabit. On this view, were an "eternal perceiver" to exist in our universe, it would be hamstrung by a temporal perspective that actually limits some if its possibilities for action.

A further consequence is that SR reveals an important enabling condition for agency. Rather than an accidental feature, the successive experience of the temporally extended agent turns out to be a *sine qua non*. We can then move from seeing temporality as a mere fact of life to a perspective that sees it as a prerequisite for thriving in our universe.

It may also be that SR actually extends further. Although I will focus on the Succession Requirement for agents, I think it is likely that SR is a precondition for selfhood as well. Unfortunately, there are nearly as many definitions of the self as there are selves, so I will not pursue this project in great depth here. However, due

---

[1] See Grondin (2010) for a scientific review.

[2] Although I am speaking of causality as a feature of our world, this should not be interpreted as a claim that "causal laws" exist independently of regularities in nature.

to the close connection with the current topic, I will take some time in this chapter to argue that J. David Velleman's (2015, p. 187) suggestion, inspired by Buddhist thinking, that we reject the enduring self and see human experience as consisting of "successive momentary subjects, each timelessly entrenched in its own temporal perspective," may be off the mark for precisely the reason that such a move invokes a plethora of non-successive experiencers, none of which can be the agents we take ourselves to be.

In making the case for SR, I will first elaborate on the concept of succession (Section 1) and then describe SR's philosophical precursors (Section 2). Next, I will show that certain apparently radical variations in temporal phenomenology collapse either into non-successive experience or successive experience, and that non-successive experiences of the kind attributed to God in the medieval period and appearing in certain popular works of fiction represent the sort of counterexamples that must be headed off in order for SR to hold (Section 3). I will then argue that non-successive experiencers (NSEs) cannot be agents because they (1) lack the appropriate causal relationships for agency, (2) exhibit the wrong sort of epistemic status for agency, and (3) lack the requisite agentive mental attitude of intentionality. Because agency is not compatible with non-successive experience, the succession requirement must therefore obtain for agents in this world.

3.1 Experiential Succession

As long as an experience is sequentially ordered, we can say an experience is *successively structured*. One kind of successive structure is quite familiar: first one thing is experienced as happening, then another thing, and then another, and so on. Another kind of successive structure might be the following: first one thing is experienced as happening, then another thing, then the first thing is experienced again, then another thing, then the second thing is experienced again, and so on.[3] Or we could reverse this order.[4] These latter possibilities are still successive in the sense of being experiences that precede and succeed one another, even if things might occur

[3] Strange experiential sequences like these are on display in the Kurt Vonnegut novels *Slaughterhouse-Five* (1969/2000), whose protagonist becomes "unstuck in time," living through temporally jumbled episodes, and *Timequake* (1997), in which the inhabitants of Earth in 2001 are transported back to 1991 to live the decade all over again exactly as it took place.

[4] The many works of speculative fiction concerning backwards succession, such as Philip K Dick's *Counter-Clock World* (1967/2012), Martin Amis' *Time's Arrow* (1991/2003), and more recently Christopher Nolan's film Tenet (2020), employ narrative structures that illustrate why "reverse" experience does not constitute a counterexample to the succession requirement. In these works, experiences of backwards succession are by necessity shown to us through successions of progressively further past sequences of events. They seem backwards to us because causation is perceived by the characters to operate in reverse, but still there is one event after another (e.g., it may be the egg goes up from the frying pan into its shell instead of vice versa, but this is an abnormality not of successive ordering but of *what* is so ordered).

in an unusual order. It is also not necessary that that which is successive be discrete in order for there to be succession (think of waves for example). I take it that everything I say here is therefore compatible with a thoroughly processual understanding of experience, despite speaking in terms which may sound superficially atomistic (e.g., speaking of a succession of entities).

It is customary in such discussions to point out another of James' oft-quoted statements, viz. "a succession of feelings, in and of itself, is not a feeling of succession," (James, 1890, pp. 628–629; cf. Hoerl, 2013 for critical analysis). We can grant this received wisdom here because SR is meant to be a necessary condition for agency and is not being posited as a sufficient condition for the temporal phenomenology of succession. SR is concerned with whether the experiences of agents must *in fact* occur successively and not primarily with whether they are felt to be successive.[5] Nonetheless, even if James is right, it is still plausible that SR is necessary for the feeling of succession, though not sufficient. I would also maintain it remains an open question whether or not, as a contingent matter, succession in and of itself *happens to be* sufficient for the experience of succession, even if it is not necessarily so.

As indicated above, the experienced succession need not be an expected ordering. What is experienced also need not be a sequence of *different* things. The content of experience can remain the same, while the structure remains successive. Identical experiences had at different times can be seen as analogous to otherwise identical things lined up, occupying different locations in space; in both cases, a structural order is maintained despite the constituents being identical (though not identical with respect to temporal or spatial position respectively).

For experiences to be successive, it is furthermore not required that the ordering of our experiences corresponds to any kind of veridical ordering of worldly events. The most well-known everyday example of non-correspondence is in dreams, which are still experienced sequentially even if the dream-events themselves seem to have a nonsensical order corresponding to nothing in particular. There are also cases of illusions that suggest we experience non-veridical event ordering in our waking lives. These include postdictive effects, in which we can seem to perceive events in a different order than that in which they actually happen.

What would it take then, for experience to be non-successive? In contrast to SR, I define Non-Succession (NS) as the proposition that agentive subjects need not have successive experiences. Non-succession is just a case where experience lacks a sequential order. In this scenario, no experiential events precede or follow one another, so there is no linear organization to experience. Alien as this may sound, claims abound alleging such experiences. It has, for instance, become a cliché in writing about life-and-death situations to describe the experience as one where "time stands still." A study by Wittmann et al. (2017) found that 120 out of 196 (61%) of assessed reports of near-death experiences gathered from a public

[5] C.D. Broad contrasted the feeling of succession, e.g., looking at a clock's second hand, and actual succession, e.g., looking at an hour hand, which, though exhibiting successive positions, is not immediately felt to be successive (Broad, 1923, p. 351).

repository mentioned a "feeling of timelessness."[6] Free-divers (Luecke, 2020),[7] users of hallucinogenic substances, and adept meditators have all reported "timeless" experiences as well (Wittmann, 2018, pp. 63–74), with proficient meditators in one study claiming they had achieved states wherein "time lost its linearity," "there was no time," or they "felt outside time and space," (Berkovich-Ohana et al., 2013, p. 912).

A common rejoinder is that this reported "timelessness" is not to be taken literally. Instead, when scrutinized, such experiences turn out to be cases of extreme time dilation (see, e.g., Arstila, 2012, on time dilation during life-threatening accidents). However we make sense of such reports, one upshot of SR is that these claims *can't* be both literal and true if they purport to involve non-successive experience.

3.2 A Brief History of the Succession Requirement

Philosophers have occasionally gestured in the direction of a Succession Requirement in the past but have stopped short of expressly arguing for it. John Stuart Mill, for instance, observed that "Sensations exist before and after one another. This is as much a primordial fact as sensation itself," (Mill, 1865, p. 198). While Mill does not explicitly contend that successively ordered sensations are a primordial fact of sensations for all agents that have them, human or otherwise, the Succession Requirement seems to lurk nearby.

We can see something of this idea in the work of earlier empiricists as well. For John Locke, succession and duration are derived from experience, much like other ideas. Locke thought that "Reflection on [the] appearances of several ideas one after another in our minds, is that which furnishes us with the idea of SUCCESSION: and the distance between any parts of that succession, or between the appearance of any two ideas in our minds, is that we call DURATION," (Locke 1694, I.xiv.3/p. 239). On the Lockean view, our memory of the actual "train of ideas which constantly succeed one another" thus forms the basis for an inference, following reflection, to the idea of succession (ibid.).[8]

Hume took Locke's idea, which anticipated James' "stream of consciousness," and applied it in a more radical way. According to Hume's "bundle theory," minds are "nothing but a bundle or collection of different perceptions, which succeed each other with an inconceivable rapidity, and are in a perpetual flux and movement,"

[6] These proportions are roughly consistent with earlier studies conducted by Bruce Grayson on the same topic, which found 74% (Greyson, 1990) and 64% (Greyson, 1983) of subjects reporting something to the effect of "time stopped" during their near-death experience.

[7] For example, one freediver interviewed by Suraiya Luecke reflected that "[The dive] does feel like somehow it happened outside the kind of linear passage of conscious time," (Luecke, 2020, Supplemental Appendix 6).

[8] See Bardon (2019) for further discussion.

(Hume 1740, 1.iv.6). Shaun Gallagher (1998) observes that, for Hume, "It is not simply that there is a stream of ideas *in* the mind; rather, the stream of ideas *is* the mind," (Gallagher, 1998, p. 9). If this is correct, then without such a succession, we do not have minds. If having a mind and being a subject amount to the same thing, then, on such a view of experience, the Succession Requirement holds. However, in keeping with Hume's antipathy towards metaphysics, he does not argue this characterization of minds is an *a priori* necessity for agency or subjectivity. Hume would instead be content with the descriptive claim that impressions indicate this streaming succession just is the case for our minds.

Inspired by Hume, Kant took a revolutionary position. For Kant, succession is not an idea derived from or simply given in experience, but rather the concept of time is an *a priori* necessity for experience, i.e., the condition for the appearing of appearances in the first place (Kant, 1787, B46). It is tempting to read Kant's "concept of time" as the necessarily successive structure of experience, but this is not clear enough to confidently assert. However, Kant does say that "time" is an invariant "form" of intuition, and that this form is necessary for experience. In Kant's system, the forms—space and time—are the fundamental grounds for experiences. Without these forms to order phenomena, distinguishing between particulars becomes impossible (Strawson, 1966, p. 49). Kant held not only that time was *necessary* for experience, but that time is *only* the scaffolding of phenomena, without discernible mind-independent existence as a thing in itself.

Kant's work does not represent ideal support for the Succession Requirement, even if it may appear he is getting at a similar contention. For one thing, Kant is not concerned with agency *per se*. Secondly, Kant's position on temporality is deeply embedded in his transcendental idealist system. In claiming that successive experience is necessary for agents, we need not also endorse the claim that time is only an *a priori* pure intuition or "form," as part of a larger view with all its attendant innovations. Finally, the arguments Kant presents on this topic are brief and not terribly conclusive (Kant, 1787, B46–B47). Some further justification for the Succession Requirement is needed.

3.3 Non-Successive Experience: The Eternal Thought Experiment

The view that *non-successive* experiences are possible has an ancient pedigree, dating back at least to Augustine (354–430 CE) and Boethius (c. 480–525 CE). Although Augustine had previously postulated that God's apprehension is non-successive in character (*Confessions,* Book XI), Boethius introduced the now classic conception of God's existence as a *totum simul*, i.e. as "everything simultaneously/all at once," (Thiselton, 2005, p. 38). According to Boethius, God enjoys "the complete possession all at once of an illimitable life," (*The Consolation of Philosophy*, Book V, Prose VI).

Henry Chadwick (1981), in his study of Boethius, interprets the doctrine of the *totum simul* as a contrast with our limited human experience: "For us, events fall into past, present, and future time. God is outside time. For him the knowledge of temporal events is an eternal knowledge in the sense that all is simultaneous present," (ibid., p. 246). In other words, God is always aware of all things that ever happen. Such a view became the standard medieval understanding after Thomas Aquinas (1225–74 CE) proposed a similar interpretation in his *Summa Theologiae*, which proposed God is immeasurable, without present, past or future, and "simultaneously whole" (Ia, Qq. 10, art. 2).

One might think there is some ambiguity in this medieval and pre-medieval understanding of God's temporality. Should we interpret Boethius as saying that, unlike humans, for God there is no privileged present, but events still retain some kind of order? I would argue this is not, or at least *not only*, the contention of the doctrine of the *totum simul*. Far from only *existing* over an infinite span of time that includes the event ordering with which we are acquainted, statements from Boethius about God's experience such as "all is simultaneous present" indicate God *experiences* everything at once, rather than in successive fashion. From the point of view of the subjective being of God, all is one experience, rendering the sequential ordering of events moot from the divine perspective. God's experience comes down to a single non-successive experience comprehending the infinite at once without a subjective distinction in order. There is no sense of one thing occurring *and then* another thing occurring for such a being. To say this total comprehension is *of* an ordered sequence is already to impose temporal order upon a subjective experience that supposedly lacks it.

The Boethian understanding of God's mode of eternal existence helps to accommodate God's omniscience, omnipresence, and temporal boundlessness (as the originator of time). However, Aquinas recognized that such a view entails God's incapacity to change (Ia, Qq. 10, art. 2). This incapacity appears inconsistent with the conception of an acting, omnipotent supreme being. As Anthony Thiselton puts it, "if eternity denotes *totum simul*, might this not be understood to impose a static mode of being onto God, who then cannot act, or interact, purposively as a living and promissory God?" (Thiselton, 2005, p. 77). Although Aquinas accepted that God's perfection required changelessness, it seems as though admitting this property limits God's ability to intervene in a way that could be understood as properly agential. We shall see in the next section that this worry is justified.

Recently there have been more secular illustrations of non-successive experience. In Ted Chiang's (2002) short story "Story of Your Life," and its 2016 film adaptation *Arrival*, earthlings are confronted by bizarre seven-limbed creatures called heptapods that experience all the events in their lives at once. These creatures employ a strange non-sequential written language called Heptapod B. The human linguist tasked with translating this language eventually learns how it is used and becomes able, like the heptapods themselves, to experience her "past and future all at once," (Chiang, 2002, p. 167). As she explains it, "I perceive—during those glimpses—that entire epoch [the 50 years of her life] as a simultaneity," (ibid.).

We can call such hypothetical subjects that experience time "all at once" non-successive experiencers (NSEs). The peculiar club of NSEs counts not only Chiang's heptapods but also previously mentioned notables like the God of Augustine, Boethius, and Aquinas, as members. The kind of subjectivity in *Story of Your Life*, though ascribed to extraterrestrials, is ultimately of the same form as the *totum simul* encountered earlier, though restricted to a mortal life.

One might think it is possible for entities to experience what we experience as time in a similar manner to the way we experience space. As Barry Dainton explains, "the fact that *we* don't perceive time itself doesn't mean that time is necessarily unperceivable: there may be logically possible worlds where space-time is substantival, as easily perceived by its inhabitants as any other part of the material furniture of their world. But our universe is not of this kind," (Dainton, 2011, p. 385). There is nothing problematic here for SR so long as these subjects experience our fourth dimension atemporally in addition to experiencing through a higher temporal dimension of meta-time that allows for succession.[9]

Kurt Vonnegut (1969) imagined just such meta-experiencers—Tralfamadorians—in his novel *Slaughterhouse-Five*. These Tralfamadorians, actually ancient hyper-intelligent robots, are described as existing outside of time, and able to "see in four dimensions," (Vonnegut, 1969, p. 26).[10] As Vonnegut describes them, "The Tralfamadorians can look at the different moments just the way we can look at a stretch of the Rocky Mountains," (Vonnegut, 1969, p. 23). This means Tralfamadorians see humans "as great millipedes—'with babies' legs at one end and old people's legs at the other," (ibid., p. 75). Hypothetical entities like these experience our familiar four dimensions through a further dimension that allows for successive experience. As such, they do not violate SR the way heptapods do and would not constitute effective counterexamples.

3.4 Against Non-Successive Experience

Following Anscombe, agency has commonly been thought to involve acting "intentionally," where intending is often specified as having a "reason for acting" (Anscombe, 1957, p. 9). Furthermore, according to standard "event-causal" theories of agency following Davidson (1963), an act's being intentional, i.e., its having

[9] We can imagine the relation between ourselves and such higher-dimensional beings would be like that between the inhabitants of "Flatland" (a two-dimensional world), and ourselves. If a three-dimensional object like a sphere passes through Flatland, the inhabitants only ever see a growing, then shrinking, two-dimensional circle. The concept of Flatland was introduced in a satirical novel by Edwin Abbott (1884).

[10] Vonnegut's (1959/1999) novel *The Sirens of Titan* also features Tralfamadorians. In that novel, the Tralfamadorian has an ordered, successive experience like ours, while his human friend Winston, who got himself stuck in a "chrono-synclastic infundibulum," experiences time all at once like heptapod.

been done for a reason, means that there is a certain kind of causal relationship between the agent's mental states (e.g., desires and beliefs) and actions.[11] The right sort of causal relationship for agency is one where the agent's intentions are causally connected to its actions, which are meant to bring about a particular effect.

If something like this account of agency is on the right track, then several serious challenges to the possibility of non-successively experiencing agents become apparent. Specifically, I will argue intractable setbacks for NSE agents arise from a faulty causal relationship between the purported agent (the NSE) and the events the agent is meant to cause. Beyond this, NSEs also lack the epistemic status we typically associate with agents. Furthermore, because of their stipulated non-successive mode of experience, NSEs lack the requisite mental states, like intentions, necessary to ground agency. These three problems together suggest NSEs would be incapable of agency given the nature of the world we live in. If non-successive experience is indeed impossible, then we should conclude the Succession Requirement must hold for agents in our world.

3.4.1 Causal Distance and Direction

NSEs face two major obstacles to agency stemming from causality: the problem of causal distance and the problem of causal direction. I will first address the problem of causal distance. Consider a hypothetical non-successively experiencing entity named Abbott.[12] Abbott, like us, is temporally extended. Only the part of Abbott temporally adjacent to an event such that a causal connection can be made between the organism and the event is capable of directly experiencing it.[13] The various temporal parts of Abbott thus could enjoy experiences of each time *at* each time, but there is no way of integrating the experiences had *at* each time as a diachronic unity spanning Abbott's life. This is because a causal connection between the parts of

[11] Schlosser (2019) helpfully distinguishes between the uncontroversial standard *conception* of agency as "the capacity to act intentionally" and the standard *theory* of agency, which fleshes out intentions with a causal account. It is not the aim of this section to argue in favor of the standard conception or standard theory, which has been done elsewhere (e.g., Davidson, 1963), but rather to investigate what these require.

[12] Abbott is a heptapod in *Arrival* (2016) and also, serendipitously, the author of *Flatland* (1884). In the film, Abbott is paired with a heptapod companion named "Costello".

[13] The notion that causes and effects must be temporally adjacent has been widespread since at least Hume. Hume's view of the causal chain has it that, "whatever objects are considered as causes and effects are contiguous; and […] nothing can operate in a time or place which is ever so little removed from those of its existence," (1740, I.iii.2). Cf. Brand (1980), Huemer and Kovitz (2003), and Mumford and Anjum (2011), who argue that simultaneous causation is possible, and Maslen (2018) for why it isn't. Regardless of whether simultaneous causation is possible at the point of direct causal interaction, it is not physically possible to have an *experience* that is simultaneously or immediately caused by environmental events because of signal transmission. Nonetheless, events and their perceptions should be causally connected, albeit mediately.

Abbott that are temporally distant from the experienced event and the experiencing, temporally proximate parts of Abbott cannot be made such that a unified consciousness over both temporal parts might obtain. This kind of connection would require that events exhibit unmediated and temporally bidirectional causal efficacy, regardless of the temporal distance between the events. However, such "spooky action at a temporal distance" is generally seen as physically impossible.[14] Short of adding a further dimension to bridge the gap, and thus allowing for a kind of succession in this new dimension, it is difficult to see how the requisite causal circumstances for diachronically unified consciousness over Abbott's lifetime would be possible in the world as we know it.

To see why Abbott would only have the relevant experiences of a time *at the time that the experienced event takes place*, it is helpful to consider the eternalist metaphysical picture of time more closely. An NSE is stipulated to exist across and to experience multiple times together, but this would not be possible if only the present exists (as presentism maintains). The eternalist position that we live in a block universe does allow for NSEs, however, as they can then spread out over and experience different times in a fully real sense.

In a four-dimensional block universe, time is not like a landscape to be traversed (Williams, 1951, p. 470). It *takes time* to perceive, let alone traverse, a landscape. A traversal of time that skips directly over certain sections necessitates another dimension to allow for the process of traversal. Sticking to a four-dimensional metaphysical picture, the experiencing being, which cannot avail itself of illicit "temporal skipping," needs to be *at a particular time* to experience events at that particular time directly (i.e., not through memories, imagination, or simulation). However, most of the temporal parts of an NSE are too causally distal to the events the NSE is meant to experience. These parts cannot therefore comprise a diachronically unified subject with one integrated experience.

Besides temporal proximity to causes, we should also expect an agent to be embedded in causal chains in the right way. If an agent were not so embedded and the subject were not affected by worldly causes, its perceptions would be of no use to guide adaptive action, as they would not carry information about the environment. The subject should therefore be embedded such that the effects of the environment arrive successively and then the subject can respond *in light* of those effects. There is thus a practical impetus to experience things successively: appropriate, informed action is enabled by the sequence of information flowing in from the environment. If part of the function of subjectivity is to integrate disparate information

[14] Cf. Adlam (2018), who argues against what she calls the "almost ubiquitous" assumption of temporal locality among contemporary physicists. However, Adlam is not explicitly concerned with causality nor with macroscopic physical processes and, as Ismael (2016, p. 132) has argued, these are the relevant physical processes with respect to agency (Adlam, 2024, has also recently given an account of "quantum agency" in the course of arguing that consciousness should be seen as inherently agential).

about the world and our bodies to adaptively guide behavior, then experiencing things in sequence is the temporal mode best suited for that function.

Causal direction poses another problem for NSEs. Consider a human translocated to a world where the temporal direction is reversed. The subject's perceptual apparatus would not be able to pick up signals from the world. The direction of causation would make subjects from our world "blind" to the reversed world and vice-versa. This is a problem for contact between subjects or worlds with differing temporal directions,[15] but, for similar reasons, it is also problematic for contact between subjects or worlds with a temporal direction like ours and subjects or worlds without a temporal direction at all. If NSEs are understood as not obeying the normal causal directionality of our world, then they would be incapable of meaningfully acting in it. Although this problem fades if we accept that, physically speaking, NSEs follow the normal causal direction, new problems then arise.

Our familiar causal direction turns out to be rooted in the physical nature of our universe, which exhibits causal asymmetry. Craig Callender, following David Lewis, has framed causal asymmetry in terms of counterfactual dependence: "Future outcomes depend upon actions now whereas past outcomes do not," (Callender, 2017, p. 259). This causal asymmetry is not just an anthropocentric illusion. As Jennan Ismael has argued, when we assess effects of interventions in the physical world from a macroscopic perspective, an emergent pattern of temporal asymmetry arises which is not present in fundamental laws of physics, but which ensures that the "direction of determination" is from past to future and not vice versa (Ismael, 2016, p. 134). As Ismael puts it, "local macroscopic interventions of the kind that correspond to visible human actions affect the macroscopic future but leave the past visibly unaffected," (Ismael, 2016, p. 132). We can then "exploit" these real temporal asymmetries in the activity of the physical components of the universe (which we can call "causal direction") to act appropriately for our own ends (Ismael, 2016, p. 129). The kind of regularities that we exploit are not mere correlations but related to, as Ismael says, "asymmetries in the way that information propagates from past to future," which are ultimately grounded in the thermodynamic gradient, i.e., the gradual increase in entropy over time in the physical universe (Ismael, 2016, p. 135; 142).

For an NSE, subjective time is stipulated not to have a particular direction, because all times are experienced at once, as opposed to successively. Experienced time would therefore not "flow" forwards or backwards. However, causation, understood as above, remains a feature of the world they and we are supposed to share. The perceptual peculiarities of the hypothetical NSE would still be subject to the direction of causation, underwritten by macroscopic temporal asymmetries resulting from the thermodynamic gradient. While for us, events in the world apparently *cause* our perceptions of events, in an atemporal world or for a purportedly

[15] See Norman Swartz (1973) on the Temporal Ozma Problem for the impossibility of contact between temporally divergent beings.

atemporal being, perceptual experience cannot be so caused.[16] The subjective experience of the NSE is explicitly atemporal in that it is detached from the ordinary direction of determination, but, like us, their physical existence inevitably succumbs to the familiar causal flow, precluding their stipulated atemporality.

If we think that experiencing a lifetime "all-at-once" means experiencing a range of times at once, then NSEs are confronted by the causal problems outlined above. However, if we think the "all-at-once" experience of an NSE refers instead to a single "static" experiential state that is itself extended over time, then one might think some causal worries can be avoided. Such a state would remain entirely changeless through time, while maintaining causal efficacy at any given time and being the effect of worldly causes at any given time. The latter kind of entity, call it a diachronic-state NSE, is not what our literary examples had in mind, but still represents an important possibility to defuse. Furthermore, one might think we can avoid causal problems by taking the radical step of shrinking our NSE down to the mysterious world of "spooky" quantum interactions, disregarding that agency may well be an emergent macroscopic phenomenon. However, neither move can avoid the further problems with which any kind of NSE must contend, to say nothing of their *prima facie* implausibility.

3.4.2 Epistemic Asymmetry

In our lives we are confronted by an apparently unavoidable epistemic asymmetry between past and future, in that we have knowledge of the past, but the future is opaque (Ismael, 2016, Ch. 6; Callender, 2017, p. 260). NSEs, however, are stipulated to exhibit epistemic symmetry. I will argue that one result of NSEs' unusual epistemic condition is that they cannot be said to make *choices*. If being able to make choices is a necessary capacity for agency, then NSEs are not agents on this count. I will also argue that enjoying such epistemic symmetry is impossible in our world anyway because epistemic asymmetry, much like causal asymmetry, is grounded in the physical structure of the universe. Far from a shortcoming, it is precisely this epistemic asymmetry that *enables* us to act in ways associated with agency. These considerations lead to the counterintuitive conclusion that NSEs, including the medieval Christian conception of God, would actually be *limited* in what they are able to do in fairly significant and startling ways. The ways in which NSEs are limited include being unable to deliberate, predict, plan, attempt, and choose, among other abilities that rely on the epistemic asymmetry we take for granted. We might think of this kind of limitation as the agentive equivalent of a car

[16] To illustrate, consider an NSE's experience at t_1 of a later event at t_5. At a minimum this would require the later event at t_5 be causally connected to the experience at t_1, but as we have seen, this is nomologically impossible. This problem confronts the NSE at any given time, because at any given time they are stipulated to experience all other times.

that cannot go any slower than 100 mph, and as such isn't very well suited to our ordinary circumstances.

In the context of agency, we can say a choice is the result of a decision process selecting between multiple events to make one or more of them come about (Ismael, 2016, p. 152). Making choices is an ability we like to believe we exercise routinely (though of course not all of our actions are choices, e.g., reflexes). An illustrative example is investing. Investment choices are typically grounded in inductive inferences, predicting future performance in light of what is known about the past. Now let's say that Abbott, as an enterprising heptapod, opens an investment firm. Needless to say, this would be the biggest money-maker on Wall Street. Abbott is, after all, privy to the future to the same extent we are to the past, as he enjoys epistemic symmetry with respect to time. In fact, Abbott's situation is *even better*—whereas we rely on corruptible records and fallible memories for information about the past, Abbott can immediately perceive every historical and future price at any given time. This is clearly a recipe for financial success.

However, Abbott is unable to make investment *choices* because he does not make a decision in order to bring about a later event. As far as Abbott is concerned, the investment is already, at any given time, perceived as invested. Or, at least, that's the nearest description we can give for Abbott's fundamentally tenseless experience. There is not even a sense for Abbott that he *will* (or *did*) invest in a particular asset as, from his perspective, it would always be as if he is in the moment of buying the asset, forever catching himself in the midst of high finance. We can see that, because of this untensed mode of experience, no process of deliberation takes place for Abbott—no prediction, no decision making, no attempt to bring about an event, and no selection process. These actions all presuppose the utilization of past information in service of enacting future outcomes in an unrealized future. Actions of this type are not only rendered irrelevant by epistemic symmetry but would be impossible except as charade.

This is not merely a problem of omniscience. Knowing how things are going to play out does not necessarily preclude choice. For instance, a human investor might have insider information such that they know tomorrow exactly what will happen, and they may know that they will capitalize on this information by purchasing a certain asset, but nonetheless *they* will still choose to buy it.

Abbott is in a different predicament. Abbot is not just omniscient but temporally *omniceptive* (to coin a term), meaning he perceives every time at once. Experiencing all times immediately is a stronger epistemic position than just knowing what happens, because experiential knowledge outstrips ordinary propositional knowledge. For Abbott it would be as if, at any point in time, he sees his portfolio including everything his investment company has ever bought, is buying, or will buy. With the entire temporal extension of the portfolio immediately transparent and immutable, there are no choices available for Abbott to make. If making choices is essential for agency, we can see then that NSEs fail to satisfy this condition. So, while Abbott's foray into finance might be exceedingly lucrative, it would not be empowering.

An NSE also does not *try* to do things. Trying requires a desire for an eventuality to come about in addition to the possibility it does not. However, from Abbott's perspective, trying would be nonsensical. Abbot might for example join an expedition to the South Pole, but he will not *try* to arrive there because, from his perspective, he is always already there (or not—he might expire *en route*, in which case, for him, he is already dying in Antarctica from the time he comes into existence).

Likewise, NSEs have no need to deliberate or plan for the future, as they are supposed to experience what we call the future as if it were what we call the present, making planning redundant (although Abbott could of course make a show of assessing an earnings report, going through the motions of an investment decision and fooling his clients). Indeed, NSEs *cannot* deliberate or plan because it is impossible for them to perceive an open future. From an NSE's perspective, there is simply nothing to plan *for*.[17]

It is likely that NSEs lack many more capacities often associated with agency that rely on epistemic asymmetry as well. One of these is the ability to "temporally decenter," or take a temporal perspective that is different from one's own (McCormack & Hoerl, 2008, p. 174). Callender, as well as L. A. Paul, maintain this temporal decentering ability is crucial to the emergence of the self, understood here as a "narrative center of gravity" (following Dennett, 1993, p. 418), and the agential perspective (Callender, 2017, pp. 255–259; Paul, 2017, pp. 263–264). Reflection on an NSE's circumstances reveals they would be deficient in this regard. For a being that experiences its entire life at once, there can be no question of taking different temporal perspectives because that being takes all of its possible temporal perspectives at any given time. There is no picking and choosing moments of its life to experience. The heptapod cannot revisit its "past" or imagine its "future." Being able to do these things requires successive experiences.

A related issue connected to the NSE's "closed future" is that, as Callender points out, "All of our evidence is confirmatory of the idea that our decision is the causal trigger that leads to, or brings about, the event. Anything prior to that decision can be trumped by the decision itself. [...] Part of what it is to be an agent is to have this sense of freedom, a sense that other future options are in some sense live," (Callender, 2017, p. 260). This same feeling does not apply to our past, about which we cannot change our minds and expect a different outcome. However, for an NSE there is no special epistemic status for the future. It is experienced like any other time. It is therefore plausible that, for an NSE, there would not be a *sense* of agency either.[18]

---

[17] See also Ismael (2016, pp. 144–153) concerning our open future and the practical asymmetry of choice.

[18] A further asymmetry that may be relevant to our sense of agency, and which NSE's lack, is the attitudinal asymmetry between past, present, and future events, or what Callender has called the "Temporal Value Asymmetry" (Callender, 2017, p. 264). This asymmetry refers to the difference in our attitudes towards and evaluations of events depending on our temporal relation to those events. So, while we might sensibly think to ourselves "Thank goodness that's over!" after surviving a harrowing Antarctic expedition, an NSE would not say the same (see also Prior, 1959, who put our temporal value asymmetry to work against eternalism).

A small wrinkle appears in these epistemic considerations, however, if a finite NSE's action takes place beyond the scope of its experience, e.g., after death. Infinite NSEs, like the medieval Christian God, would be omniceptive over eternity, but finite NSEs, like heptapods, could still be ignorant of those things which take place after their life ends. Thus, regarding circumstances beyond the scope of their total experience, an NSE runs into the same epistemic asymmetry we do and consequently may appear more like an agent, albeit from beyond the grave.

Even so, there are still reasons for thinking any NSE's epistemic setup just wouldn't work in our universe. An NSE faces no pressure to gather information to guide its action and yet it is this task our minds appear to have evolved specifically to carry out. The reasons we face an epistemic asymmetry are also not an accidental feature of our human minds. Once again, there are physical facts constraining us. As Ismael puts it, "what explains our greater knowledge of the past than the future is that along [the thermodynamic] gradient, inferences from the present, surveyable macroscopic state of the world to its past [...] are much more powerful inferences than inferences from present to future," (Ismael, 2016, p. 143). Ismael goes on, "the thermodynamic gradient makes it possible to create records in the environment that will carry information to our future selves," (Ismael, 2016, p. 145). What the gradient *does not do* is make it possible to carry information from our future selves to our past selves, which is what an NSE needs to do to exist as a diachronically unified entity. To us, the future is not only epistemically inaccessible but seems as if it has not yet occurred and will come to be shaped by us. The propagation and accumulation of information apparently works in one direction, such that even if the future already exists in some sense, *it might as well not* to any being in our universe. This is a potentially insurmountable obstacle to the actual existence of an NSE. An upshot of this is that our own successive mode of experience should be seen not as limitation, but on the contrary, an enabler of a kind of existence that is uniquely suited to thrive in the world we live in.

3.4.3 Intentionality

As mentioned previously, on standard accounts of agency, agents need to have particular kinds of mental states. Specifically, agents need to be able to form intentions and act on them, and these intentions should be causally related to the eventual actions. I will now argue non-succession precludes the formation of intentions, thus ruling out agency as traditionally understood. NSEs should be seen as incapable of forming intentions for two main reasons. One is a fairly straightforward worry related to the problem of epistemic asymmetry. The worry is that, while we intend future events to come about as a result of our actions, the future is not available to an NSE as a domain of uncertainty. It then does not make sense for an NSE to *intend* that future events come about, as these events have no special status and are experienced directly. Just as we do not form intentions concerning the past, NSEs would not form intentions about the future.

A second, more troublesome worry is rooted in the ontology of mental states like intentions, itself intimately related to the ontology of experiences. Experiences are not static, unchanging entities, but evolving processes defined, at least in part, by their dynamism. To make this thought more ontologically explicit, we can follow O'Shaughnessy (2000, Part I.2) and Soteriou (2013, Ch. 6), who claim experience is "not merely a *continuous existent* across time, [but] an activity and therefore also a *process*, and thus occurrently renewed in each instant in which it continues to exist," (O'Shaughnessy, 2000, p. 42).[19] Occurrent processes are defined by being *successively present* (Soteriou, 2010, p. 139). Unfolding processes, as opposed to states, also allow experienced events to have duration, i.e., to take time. If we agree that experience should be defined in this way, then an NSE is incapable of experience as such. This is because, for a non-successively experiencing being, every event of its life is supposed to be experienced all at once at each moment. There can then be no individuatable experiences at different times, separately caused by different events or that cause different events at different times. Instead, there would be just one state.

While the above calls into question the very possibility of non-successive experience, one still might think a non-experiencing entity could be an agent. It is perhaps still possible with this understanding that NSEs have intentions, provided *intentions* are not necessarily processive. If an intention is just a set of dispositional facts, for instance, this could be the case. Abbott could still exhibit an innate set of non-qualitatively felt dispositions that occupy a causal role(s). However, such dispositional properties do not seem distinctive of those entities we consider to be agents, nor relevant to their status as such. Inanimate objects like toasters also have dispositions to behave in certain ways but seeing toasters as intentional agents does violence to our ordinary concept of agency (cf. Dennett, 1987).

I would argue instead that mental states like intentions must be occurrent processes. One reason for thinking this is the case is that intentions have a distinctive cognitive phenomenology—it *feels like* something to intend to do something, and it *feels like* we are the ones doing it.[20] There is an urge towards action preceding and ostensibly precipitating the action. For genuine agents, it is not just that we act because we are disposed to do so, even if dispositions play a role. Mental states like intentions are, at least in part, *experienced*, and as such, must be processive for the same reasons as experiences. Without this experiential component, the NSE lacks a

[19] See also Bartlett (2018), who argues nearly all philosophers of mind hold that experiences are processive, even if this is not explicitly stated. As I have indicated in this paragraph, the process view of experience may shut down the possibility of NSEs *a priori*, not just as agents, but in any sense. However, one can remain sceptical of the process view of experience and the arguments in this section concerning intentions should still apply.

[20] See Horgan (2011) for further discussion and defense of the cognitive phenomenology of agency.

critical element involved in our understanding of intentions, and thus also fails the intentional criterion for agents.

Beyond the inappropriate cognitive phenomenology, the changeless intentions of an NSE would be unable to play the role of cause to all the actions of the subject. In order to do so the state would have to exhibit differing causal powers and susceptibilities at different times, which it can't do without changing in some way. However, for the NSE, there can be no change in intention, as a change in intention necessitates an experiential succession.[21]

Now, it may appear this problem is avoided if a mental state can change without the NSE's experience changing. So long as we accept there are subpersonal mental states such a contention seems eminently plausible. However, the *kinds* of mental states that are most relevant to agency are not subpersonal ones. It is personal-level mental states that must have the appropriate causal role, otherwise we are hard pressed to distinguish automatic reactions from the actions of an agentive subject. As a result of these considerations, we should think NSEs are incapable of forming dynamic intentions and thus also incapable of the kind of intentionality that agency requires.

One might think there may yet be ways of salvaging agency for NSEs. For instance, NSEs might have *sui generis* non-processive intention-like states that occupy the functional role of an intention without phenomenal character. These non-processive, non-experiential states might be causally related to actions and their effects. Limited to a changeless state extended over time, and correspondingly limited in causal power, without cognitive phenomenology, experience, or the various abilities predicated on epistemic asymmetry, such an entity would stray radically from our ordinary understanding of agency, and, in my view, would no longer represent a challenge to the Succession Requirement because of this.

3.4.4 NSEs and Non-Standard Theories of Agency

We might still wonder how NSEs would fare on a non-causal approach to agency (e.g., Melden, 1961; O'Connor, 2000). On such accounts, the agent need not stand in any particular causal relation with their actions. Understanding agency in this way doesn't let NSEs off the hook for several reasons. One is that the problems of causal asymmetry are not only problems for agency, but also problems for the NSE's

[21] As Merleau-Ponty puts it: "Change presupposes a certain position which I take up and from which I see things in procession before me: there are no events without someone to whom they happen and whose finite perspective is the basis of their individuality," (Merleau-Ponty, 1945, p. 477).

existence as a diachronically unified subject. There just cannot be direct causation between temporally distal events or non-directed causation on the macroscopic scale, yet these conditions must obtain in order to furnish the NSE with the experience it is stipulated to have. However, even ignoring these causal issues, an NSE would still run into epistemic asymmetry problems and the problem of non-intention.

Suppose however we agree with Anscombe (Anscombe, 1957, p. 9) that intentional actions are those which are done for a reason, but reject the standard contention that agency also requires a special causal connection between intention and events. Perhaps then, it might seem, NSEs can at least avoid the problem of non-intention. We might think a non-successively experiencing being may still have reasons to act in certain ways, even though such reasons may have to be as "ever-present" to the NSE as the NSE's experience.

I do not think such a move would be successful. Appealing to reasons requires the subject enjoy certain mental states, including belief states. However, as discussed earlier, these mental states have a distinctive cognitive phenomenology that suggests they must be processive in character and so not the kind of thing that can be had at a single moment. Instead, it is essential that beliefs are experienced successively. Perhaps more importantly, though, is that for an agent to act according to reasons, the reasons should *precede* the action. Unfortunately, non-succession makes this impossible by definition.

3.4.5 Losing Your Self in the Moment

So far, we have been considering experiential succession as a precondition for agency, but it is likely experienced succession is also a condition for selfhood. While I think this idea is somewhat intuitive, it is not universally shared. J. David Velleman, for instance, drawing on Buddhist thinking and Derek Parfit's "neo-Lockean" account of persons (Parfit, 1984, p. 281), thinks "the existence of an enduring self is an illusion," (Velleman, 2015, p. 175). For Velleman, however, the emphasis is on *enduring*. An enduring self persists through time in such a way that it is whole at any given time. Though this might be how we ordinarily conceive of ourselves, Velleman argues selves are in fact not like this at all. Rather, selves should be seen as *perduring* objects, in that they persist by the succession over time of individual temporal parts.

Recognizing ourselves as perduring entities, Velleman hypothesizes, could induce a radical and potentially liberating perspective shift. As he puts it: "Suppose that I could learn to experience my successive moments of consciousness — *now* and *now* and *now* — as successive notes in a performance with no enduring listener, no self-identical subject for whom these moments would be *now* and *then* and *then again*," (Velleman, 2015, p. 187). Your present self, in this scenario, "would think of itself, and each of the subjects with whom it communicates by memory and anticipation, as seeing its own present moment, with none of them seeing a succession of moments as present," (ibid.). This present self, and indeed each momentary self on such a conception, should by now look very familiar. What we have here is a collection of non-successive experiencers together comprising what we mistakenly think is an enduring self.

Velleman goes on to claim that for one's present self, "time would no longer seem to pass, because [one's] experience would no longer include a subject of its passage — just successive momentary subjects, each timelessly entrenched in its own temporal perspective," (ibid.). As tentative evidence for this, Velleman cites research on "flow-states" during which we tend to "lose awareness of time's passing" and lose self-awareness to boot (Csikszentmihalyi, 1990). Velleman is concerned with the extent to which we can shift our perspective towards the perduring self and whether this could console us and alleviate suffering. While Velleman ultimately admits that, as far as we ordinary humans are concerned, "we can't stop the self from seeming to endure," (Velleman, 2015, p. 192), I would argue the thought experiment itself is incoherent, insofar as it presents NSEs not only as possible but as our fundamental constituents.

We have already seen that NSEs cannot be agents. We would therefore be remiss in applying agential language to the time slices of Velleman's hypothetical scenario. The notion of self does not apply here either. If we identify the agentive self with a "locus of control," (Ismael, 2016, p. 101), then without successive experience such a self does not exist, given the arguments of previous sections. Other conceptions of the self also fail to obtain without successive experience. Consider, for example, that there are a number of essentially processive actions, such as "seeing," "thinking," and "communicating," which can't be said to be things that a *momentary* subject is ever capable of doing. There are predication errors in describing a momentary self as interacting with its environment, itself, and other subjects. The reason for the errors in predication is that the appropriate grammatical subject to which the predicates would be applied has dropped out of existence by failing to extend over time. We need that subject to exist wholly at multiple times, not only at one time, in order to predicate of that self the processive actions we typically want a self to be able to do.

A more accurate description of the consequences of fully realizing the perduring self would be an *absence* of an experiencing self. Non-successive experience is not just nomologically impossible for agents, it also entails non-selfhood. This does not mean a state of non-succession isn't possible; it may be, but only if we eliminate both selves and our agency. This sort of total "ego-death" is reported by users of hallucinogens (see, e.g., Nour et al., 2016) and is arguably the goal of some forms of meditation. Many Buddhist traditions,[22] for example, do not seek a scenario like

[22] There is great diversity within Buddhism, but the concept of "no-self" is widespread, appearing for example in the *Dhammapada*, believed by the faithful to be a record of Buddha's own words. The interpretation of this concept is naturally the subject of much debate. See Baggini (2018, pp. 175–188) for an accessible overview.

Velleman describes, with its multiplicity of selves, but instead seek "no-self" (Sanskrit: *anātman*).[23] If the truth is that we are perduring rather than enduring entities, fully realizing this truth would be more world-shattering than Velleman's thought experiment suggests.

There are potential ethical consequences here as well. If SR is true, as I have claimed, of both agency and selfhood, then it is likely NSEs would not be appropriate targets of moral judgment. Because non-agents do not have a claim to moral responsibility, it would make as much sense to call an NSE's actions "good" or "bad" as it would to describe a toaster this way. Taking the thought to its most provocative conclusion, if we consider our most famous alleged NSE—the medieval Christian God—it seems we have to say claims like "God is good" involve errors in predication as well.

It should be plain that I am not claiming selves or agents necessarily *exist* in any strong sense. What I *am* claiming is that, for such things to exist in an ontologically strong sense *or* as some sort of illusion,[24] it is paramount that we as subjects enjoy a successively structured experience. There are, of course, very many different conceptions of "self," and exploring the role that the temporal structure of our experience plays for each of these exceeds the scope of the present chapter. However, I hope this brief excursion has highlighted the potential extensibility of SR to domains beyond agency.

3.5 Conclusion

I have argued that successive experience is necessary for agency in our world. I have tried to demonstrate this by showing that non-successive experiencers are nomologically impossible. Non-successive experiencers face several intractable problems for agency. These include being too causally distant from causes and effects and being inconsistent with causal direction. In addition, non-successive experiencers lack the epistemic asymmetry associated with agency. Finally, non-successive experiencers are seemingly incapable of having mental states like intentions, which are essential to agency. After establishing the Succession Requirement for agents, I then suggested that SR can likely be extended to selfhood as well, and perhaps further.

---

[23] D. T. Suzuki, for example, quotes a poem by the Zen master Bunan: "While alive/Be a dead man,/Thoroughly dead;/And act as you will,/And all is good," which he then expounds upon: "To love God is to have no self, to be of no-mind, to become 'a dead man,' to be free from the constrictive motivations of consciousness," (Suzuki, 1986, p. 16).

[24] Dennett (1993), for example, claims the self doesn't exist except as a kind of "virtual self." For this we need to construct a serial narrative out of a jostling effervescence of mental activity. So it is that, even for the illusory self to exist, successive experience comes along with it.

In this world, it is reasonable to think that agency is impossible without succession and that agentive subjects require a successive temporal structure in their experience. We can expect, therefore, that agents we encounter in our world will enjoy the same kind of successive temporal structure that we do. Entities without a successive experiential structure, on the other hand, should not be considered genuine agents. Furthermore, I suspect such entities could not be subjects either, at least not if subjective experience is fundamentally agentive and/or processive in character.

References

Abbott, E. A. (1884). *Flatland: A romance of many dimensions*. Seeley & Co.

Adlam, E. (2018). Spooky action at a temporal distance. *Entropy, 20*(1), 41.

Adlam, E. (2024). How do we observe relational observables? *Journal for General Philosophy of Science*. Preprint. https://doi.org/10.48550/arXiv.2410.05508

Almäng, J. (2014). Tense as a feature of perceptual content. *The Journal of Philosophy, 111*(7), 361–378.

Anscombe, G. E. M. (1957/2000). *Intention*. Harvard University Press.

Arrival. (2016). *[film] Directed by D. Villeneuve*. 21 Laps Entertainment.

Arstila, V. (2012). Time slows down during accidents. *Frontiers in Psychology, 3*, 196.

Baggini, J. (2018). *How the world thinks*. Granta Books.

Bardon, A. (2019). Explaining temporal phenomenology: Hume's Extensionalism and Kant's Apriorism. *Kant-Studien, 110*(3), 463–476.

Bartlett, G. (2018). Functionalism and the problem of Occurrent states. *The Philosophical Quarterly, 68*(270), 1–20.

Berkovich-Ohana, A., Dor-Ziderman, Y., Glicksohn, J., & Goldstein, A. (2013). Alterations in the sense of time, space, and body in the mindfulness-trained brain: A neurophenomenologically-guided MEG study. *Frontiers in Psychology, 4*, 912.

Brand, M. (1980). Simultaneous causation. In P. Van Inwagen (Ed.), *Time and cause* (pp. 137–153). Reidel Publishing.

Broad, C. D. (1923). *Scientific thought: A philosophical analysis of some of its fundamental concepts*. Kegan Paul.

Callender, C. (2017). *What makes time special?* Oxford University Press.

Chadwick, H. (1981). *Boethius: The consolations of music, logic, theology, and philosophy*. Oxford University Press.

Chiang, T. (2002/2015). Story of your life. In *Stories of your life and others* (pp. 109–172). Picador.

Csikszentmihalyi, M. (1990). *Flow: The psychology of optimal experience*. Harper & Row.

Dainton, B. (2011). Time, passage and immediate experience. In C. Callender (Ed.), *Oxford handbook of philosophy of time* (pp. 382–419). Oxford University Press.

Davidson, D. (1963). Actions, reasons, and causes. *Journal of Philosophy, 60*, 685–700.

Dennett, D. (1993). *Consciousness explained*. Penguin.

Gallagher, S. (1998). *The inordinance of time*. Northwestern University Press.

Greyson, B. (1983). The near-death experience scale: Construction, reliability and validity. *Journal of Nervous and Mental Disease, 171*, 369–375.

Greyson, B. (1990). Near-death encounters with and without near-death experiences: Comparative NDE scale profiles. *Journal of Near-Death Studies, 8*, 151–161.

Grondin, S. (2010). Timing and time perception: A review of recent behavioral and neuroscience findings and theoretical directions. *Attention, Perception, & Psychophysics, 72*(3), 561–582.
Hoerl, C. (2013). A succession of feelings, in and of itself, is not a feeling of succession. *Mind, 122*(486), 373–417.
Horgan, T. (2011). From agentive phenomenology to cognitive phenomenology: A guide for the perplexed. In T. Bayne & M. Montague (Eds.), *Cognitive phenomenology* (pp. 57–78). Oxford University Press.
Huemer, M., & Kovitz, B. (2003). Causation as simultaneous and continuous. *The Philosophical Quarterly, 53*, 556–565.
Hume, D. (1740/2011). A treatise of human nature. In *Hume: The essential philosophical works*. Wordsworth.
Ismael, J. T. (2016). *How physics makes us free*. Oxford University Press.
James, W. (1890/1980). *The principles of psychology*. Henry Holt.
Kant, I. (1787/2007). *Critique of pure reason* (N. K. Smith, Trans.). Palgrave Macmillan.
Luecke, S. (2020). *The neurophenomenology of freediving: An investigation of brain, body, and behavior through breath*. MSc dissertation. University of Edinburgh.
Maslen, C. (2018). The case for widespread simultaneous causation. *The Philosophical Quarterly, 68*(270), 123–137.
McCormack, T., & Hoerl, C. (2008). Temporal Decentering and the development of temporal concepts. *Language Learning, 58*, 89–113.
Melden, A. I. (1961). *Free action*. Routledge and Kegan Paul.
Merleau-Ponty, M. (1945). *Phenomenology of perception* (Colin Smith. Trans.). Routledge.
Mill, J. S. (1865/1979). *An examination of sir William Hamilton's philosophy, and of the principle philosophical questions discussed in his writings*. University of Toronto Press.
Mumford, S., & Anjum, R. (2011). *Getting causes from powers*. Oxford University Press.
Nour, M. M., Evans, L., Nutt, D., & Carhart-Harris, R. L. (2016). Ego-dissolution and psychedelics: Validation of the ego-dissolution inventory (EDI). *Frontiers in Human Neuroscience, 10*, 269.
O'Connor, T. (2000). *Persons and causes: The metaphysics of free will*. Oxford University Press.
O'Shaughnessy, B. (2000). *Consciousness and the world*. Oxford University Press.
Parfit, D. (1984). *Reasons and persons*. Clarendon Press.
Paul, L. A. (2017). The subjectively enduring self. In I. Phillips (Ed.), *The Routledge handbook of philosophy of temporal experience* (pp. 262–271). Routledge.
Prior, A. N. (1959). Thank goodness that's over. *Philosophy, 34*(128), 12–17.
Schlosser, M. (2019). Agency. In E.N. Zalta (Ed.), *The Stanford Encyclopedia of Philosophy* (Winter 2019 Edition), Retrieved February 20, 2021, from https://plato.stanford.edu/archives/win2019/entries/agency/
Soteriou, M. (2010). Perceiving events. *Philosophical Explorations, 13*(3), 223–241.
Soteriou, M. (2013). The mind's construction: The ontology of mind and mental action. Oxford University Press.
Strawson, P. F. (1966/2007). *The bounds of sense: An essay on Kant's* critique of pure reason. Routledge.
Suzuki, D. T. (1986). *Studies in Zen*. Unwin Paperbacks.
Swartz, N. (1973). Is there a an Ozma-problem for time? *Analysis, 33*(3), 77–82.
Tenet. (2020). [film] Directed by C. Nolan. Hollywood: Syncopy.
Thiselton, A. C. (2005). *A concise encyclopedia of the philosophy of religion*. Baker Academic.
Velleman, J. D. (2015). So it goes. In J. D. Velleman (Ed.), *Beyond price: Essays on birth and death* (pp. 175–193). Open Book Publishers. Originally presented as the first Amherst Lecture in Philosophy on March 9, 2006 and published online at http://www.amherstlecture.org/velleman2006/index.html. Reprinted (2010) in J. Perry, M. Bratman, and J. M. Fischer (Eds.), Introduction to Philosophy: Classical and Contemporary Readings, 371–382.
Vonnegut, K. (1959/1999). *Sirens of titan*. Millenium.

Vonnegut, K. (1969/2000). *Slaughterhouse-five or the children's crusade: A duty-dance with death*. Vintage Classics.
Williams, D. C. (1951). The myth of passage. *The Journal of Philosophy, 48*(15), 457–472.
Wittmann, M. (2018). *Altered states of consciousness: Experiences out of time and self* (P. Hurd, Trans.). MIT Press.
Wittmann, M., Neumaier, L., Evrard, R., Weibel, A., & Schmied-Knittel, I. (2017). Subjective time distortion during near-death experiences: An analysis of reports. *Zeitschrift für Anomalistik, 17*, 309–320.

Chapter 4
The Classic Models of Time-Consciousness

We now turn from describing subjective time and isolating its most fundamental features to considering competing analyses of temporal experience. Broadly speaking, high-level models of subjective time fall into one of three categories: cinematic, retentional, or extensional. We can refer to these as the "Classic Models." The explanatory target of these models is time-consciousness, which is to say, they aim to accurately describe and analyze the temporal phenomenology of our present experience. The three classic models are typically presented in opposition to each other, with the assumption that one of them can offer the correct analysis. However, the classic models are in reality neither mutually exclusive nor jointly exhaustive, but rather represent the most prevalent contemporary and historical positions. This chapter will consider what makes these classic models distinctive and why each is ultimately unsatisfactory on its own. This chapter also defends the doctrine of the specious present, but finds that neither of the classic specious present views sufficiently accounts for time-consciousness as we find it. The next chapter (Chap. 5) will continue the defense of the specious present against a recent challenger known as the dynamic snapshot view, and in so doing will make a case against localized mechanisms underwriting time-consciousness. The following chapter (Chap. 6) will then return to the classic models and argue that we should ultimately prefer a particular kind of extensional-retentional analysis, which I am calling "ERA." Before we can understand this positive proposal, however, some stage setting in order.

Models of the experiential structure of time are supposed to illuminate or resolve questions about how our oft-paradoxical temporal phenomenology can be as it is. In other words, they attempt to answer how we can have experiences of continuity, succession, etc.—those phenomenal features that constitute theoretical explananda, which have already been discussed at some length. We can think about the various models of subjective time as resulting from different answers to questions like the following: "If we really are capable of directly apprehending temporally extended phenomena, how is this possible? What must our consciousness be like—what

C. A. McKenna, *Constructing the Present*, Studies in Brain and Mind 24,
https://doi.org/10.1007/978-3-031-89571-5_4

features and structures must it possess—for this to be possible?" (Dainton, 2014, p. 101). For so-called extensionalists, the answer to the above question is that the experience of time is *itself* extended in time. For retentionalists, the experience of time is *not* extended in time (or at least needn't be). Rather, in retentionalism, the content of a succession of intentional mental states or processes, each *representing* an extended experience, grounds the phenomenology of extended time. In contrast to both of the former views, the cinematic model has it that our experiences are analogous to frames in a movie, each a mere moment in succession without necessarily representing an extended experience. A basic schematic of the three models is provided in the Fig. 4.1:

4.1 The Specious Present

All the classic models position themselves in relation to the now orthodox[1] doctrine of the "specious present." It will be useful, therefore, to further elaborate on this concept before considering the models themselves. William James initially popularized the term "specious present,"[2] to refer to the interval of time that a subject takes to be "now," in contrast to the objective present of physics, which we can call "clock time." James' insight was that, for the subject, the present is *not* instantaneous like an "invisible knife-edge," but *lasts* a certain amount of time (James, 1890, p. 609).

While the extensionalist and retentionalist models can both be seen as different ways of accommodating the specious present, the cinematic model is usually construed as opposing the specious present doctrine. Because it only posits *instantaneous* or *momentary* experiences, like frames in a film, the cinematic model lacks an obvious mechanism for accommodating a subjective present that occupies an interval. It could be the case that some collection of these successive momentary experiences is meant to constitute an overarching subjective present beyond the experiential moment itself, but this takes away some of the dialectical force of the model, which emphasizes a succession of "nows," each of which corresponds to a moment in actual time. Potential reconciliation notwithstanding, a tradition of "cinephiles" disputing the specious present has arisen, and their objections will be the focus of the next section.

---

[1] See Kent and Wittmann (2021) on the predominance of the specious present view in the scientific literature. They claim that, despite criticism of the specious present by White (2017) and Arstila (2018), among others, the idea of a specious present is convincingly borne out by contemporary empirical studies (e.g., Lloyd, 2012; Berkovich-Ohana & Glicksohn, 2014; Northoff, 2016) and remains a standard theoretical assumption.

[2] Although James was most responsible for popularizing the term "specious present," he claims the originator was a "Mr. E. R. Clay" who is now thought not to have existed—instead, the attribution rightfully belongs to Bostonian cigar manufacturer and amateur philosopher E. Robert Kelly who published an anonymous book called, "The Alternative: A Study in Psychology" in 1882 (for more historical background see Andersen, 2014).

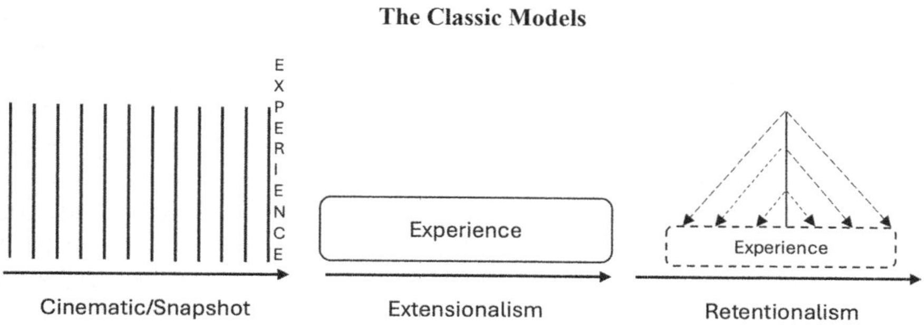

Fig. 4.1 The three classic models. The rightward pointing arrow represents clock time. The cinematic model involves a succession of momentary experiences mapping onto moments in clock time. The extensional model shows experience occupying a real interval mapping onto some duration of clock time. The retentional model shows trailing retentions and protentions (dotted arrows). In most forms of retentionalism, it is the experiential *content* of the retentional model that determines specious present, so while the experience is *as of* a duration, it is not actually extended in clock time. This *as of* is represented by the dotted border. Some retentional models do not include protentions, though Husserl's view, which is the progenitor of the type, does (Husserl, 1917). This figure was inspired by a similar diagram by Barry Dainton (2023)

Multiple considerations, both *a priori* and empirical, originally motivated the doctrine of the specious present. Most pressing was the problem of change and succession, i.e., the problem of accounting for how feelings of change and succession can arise from mere successions of moments. To overcome this problem, James popularized the view that we perceive the present not as a single infinitesimally short moment but as an interval or span with "a rearward- and a forward-looking end" (James, 1890, p. 609). For there to be an experience of change and succession, James thought there must be an interval wherein change or succession can occur because the experience of these things essentially involves a contrast of multiple states. The resultant "duration-block" constitutes the fundamental "unit of composition" of our temporal experience (ibid.). James adopted the term "specious present" for such an interval because, as a length of time, it does not correspond to the "objective" present as traditionally understood (i.e., as an infinitesimally narrow interface between the future and the past). There is thus a mismatch between what we subjectively *take to be present* and the supposed instantaneous present of the world itself.

Moreover, there must be information about past and future states contained within the present in order for us to experience things *as* fading into the past and arriving from the future. Barry Dainton has called this now widely accepted idea "The Principle of Simultaneous Awareness": we must be simultaneously aware of more than an instantaneous present for us to be able to experience a succession at all (Dainton, 2006). The word "simultaneous" is hazardous here as it does not necessarily imply subjective simultaneity, i.e., that we experience two things at the same time, but rather that more than one moment must be co-conscious or present in mind at the same time.[3] The past and future must be in some sense *contained within* the experience of the present, whether that present is momentary or otherwise, in order to give the present its distinctive phenomenal character.

So far these have been essentially *a priori* considerations. James also noted from anecdotal and experimental evidence that people really did report and act as though the present occupied an interval of time. An interval conception of the present is evident in common parlance, as Thomas Reid (Reid, 1785), and later Shadworth Hodgson, whom James quotes in a footnote, observed: "The present moment is often extended practically to a few seconds, or even minutes, beyond which we specify what length of time we mean, as the present hour, or day, or year, or century," (Hodgson, 1878, pp. 248–254). We tend to speak as if the present is a variable length of time—when we say, for instance, that someone is living in a new house now, it does not imply that they are there at this very moment, but that they inhabit this new house through some nebulous duration we conceive of as being present. Of course, linguistic quirks do not necessarily tell us anything about our phenomenology, but they at least indicate that the specious present is not such a foreign concept as it initially appears.

James also thinks little introspection is necessary to reveal that the phenomenology of temporal experience is characterized by the specious present. As he says, "We do not first feel one end and then feel the other after it, and from the perception of the succession infer an interval of time between, but we seem to feel the interval of time as a whole, with its two ends embedded in it," (James, 1890, p. 610). We *feel* as though a certain space of time is held in mind at the same time. This allows us to perceive a musical note as having succeeded from an immediately prior note, with the expectation that it's moving in the direction of a different note. The phenomenal character, or feeling, of a certain note is, on such a view, different if it occurs in isolation rather than in sequence because the character of the phenomenal present is not determined by one isolated moment but by an interval.

It should be noted that the timespan of the specious present constitutes the present of *phenomenally conscious* perception, which involves any experience where there is "something it is like" to have such an experience.[4] The specious present is *not* the time between when an event occurs and when the central nervous system is able to produce behavioral reactions. Lag in processing time can vary depending on the implicated neurons, e.g., from between 10–30 ms for some behavioral reactions independent of conscious awareness to 30–60 ms for saccadic reaction times (Frost & Pöppel, 1976). The specious present, however, is a domain specifically concerning the interval the phenomenally conscious subject takes to be the present—the

[3] James quotes Volkmann for a more representationalist spin on this same thought: "if A and B are to be represented *as occurring in succession* they must be *consciously represented*; if we are to think *of* them as one after the other, we must *think* them both at once," (Volkmann, 1884, §87).

[4] This terminology follows Nagel (1974).

present which is always a part of our experience of anything and is in that sense a constant feature of consciousness.

While experiments have helped to give us a range of possible neural processing times, it is much less clear how we should answer the question of how long the specious present is. James himself seems to indicate a range of as little as three quarters of a second (James, 1890, p. 615), corresponding to the "minimum amount of duration which we can distinctly feel," up to 12 s as "the maximum filled duration of which we can be both distinctly and immediately aware," (James, 1890, p. 613). Meanwhile, modern commentators have come up with intervals of their own: Dainton maintains the specious present would be roughly 0.5 s long (Dainton, 2006), Lockwood gives an interval of 1 to 1.5 s (Lockwood, 2007), Pöppel and Bao cite studies that converge on a 30 ms window for "primordial events," that constitute building blocks that come together within the 2–3 s window of the subjective present (Pöppel & Bao, 2014). Wittmann (2014) accepts a similar window of approximately 3 s, while Grush (2005) advocates a smaller figure of 200 ms.

However, if we remember that the specious present was initially proposed as a way of dealing with the problem of succession, then a good place to start in nailing down the length of the interval is to look at the threshold below which we can no longer experience successive events. If Dainton is right that "two knocks" can be experienced as successive when the interval between them is greater than half a second but not below, then it seems half a second is approximately accurate as a duration of the specious present (Dainton, 2011). A short timescale of such a magnitude is adequate for the explanation of succession, but as Arstila and Lloyd point out, there remain other aspects of the phenomenology of time-consciousness that can occur over longer durations, e.g., continuity in the stream of consciousness, which would then have to be accommodated by something beyond the specious present (Arstila & Lloyd, 2014a).

Many experimental attempts to specify a precise length for the specious present use reaction time paradigms. Such designs rely on the implicit hypothesis that relevant cognitive processes are sequential, or linear, forms of information processing. However, cognition involves parallel processing. The experimental design of reaction time studies unfortunately encourages the suppression of the effects of parallel processing in favor of a superficial linearity, as Pöppel and Bao describe: "the experimental setup may select one activity from these parallel operations and shift it into a frame of sequentiality; by doing so, the experimenter may be seduced into concluding that other activities that have not been selected are in fact nonexistent," (Pöppel & Bao, 2014, p. 245).

In reality there may not be a definite answer to the question of how long the specious present is, but instead only a variable, context- and task-dependent range. We should expect this interval to be subject to a high degree of modulation, much like other neural processes, contracting and dilating depending on circumstances, for example as we encounter and resolve life-threatening situations. The experience of time is likely not as unified as it appears, with "intra- and multimodal mechanisms that have different temporal characteristics" resulting in varying durations of the specious present depending on exactly which phenomena and which modalities we are investigating (Arstila & Lloyd, 2014a). Thus instead of *one* specious present there could be many, although there should still be approximate upper and lower bounds for these intervals determined by physical constraints on their underlying neural processes.

4.1.1 The Ambiguity of the Specious Present

A further source of tension among commentators on the specious present is that, from a conceptual standpoint, the doctrine as initially presented in *The Principles of Psychology* is ambiguous between at least two interpretations.[5] Extensionalist and retentionalist models can both be seen as attempts to resolve this ambiguity and accommodate the specious present in different ways, with each claiming to better capture the phenomenal character of temporal experience. Towards the beginning of James' chapter on the perception of time, it seems we are meant to take the notion of a temporal interval or duration quite seriously indeed, to the effect that we should see the present as being grounded in experiences that are themselves extended in time. James' contention that a "duration-block" comprises the fundamental unit out of which consciousness is composed seems suggestive of such an interpretation (or, as James puts it: "the original paragon and prototype of all conceived times is the specious present, the short duration of which we are immediately and incessantly sensible [i.e. aware]," (James, 1890, p. 631)). This is essentially the extensionalist reading and is the simpler of the two. Extensionalists accommodate the specious present by proposing that all experiences are essentially extended in time, occupying some minimal duration, and these intervals are successive in some way, perhaps overlapping, perhaps not, depending on the author.

However, as James continues his investigation, it becomes clear that he has a rather more sophisticated model in mind. Especially when quoting Volkmann and

[5] For more on the inherent ambiguity in James' presentation of the specious present, see Arstila and Lloyd (2014b) and Le Poidevin (2015). According to Le Poidevin, the two biggest sources of confusion are over (1) whether James thought "'the specious present' refers to the object of the experience, namely a duration in time, or the way in which that object is presented to us" and (2) at least 4 potential disambiguations of James' contention that the specious present is "immediately sensible." Le Poidevin concludes that we could define the specious present, while remaining true to James' own words, in the following four ways: "1. the span of short-term memory"; "2. the duration which is perceived, not as a duration, but as instantaneous"; "3. the duration which is directly perceived—i.e. not through the intermediary of a number of other, perhaps instantaneous perceptions"; and "4. The duration which is perceived both as present and as extended in time," (Le Poidevin, 2015). Arstila and Lloyd also see an inherent ambiguity between an extensionalist and retentionalist reading, and even a modified form of the cinematic view where the (normally instantaneous) frames themselves occupy some duration (Arstila & Lloyd, 2014a, p. 310). The latter "cinematic" interpretation is dubious, as it is not clear why we should continue to think of this as a cinematic view and not a form of extensionalism if the experience is extended, but ultimately this is a terminological dispute.

Ward, James seems to endorse a succession of representations. On this reading, at any given time, multiple representations are present and the "*content* is in a constant flux, events dawning into its forward end as fast they fade out of its rearward one, and each of them changing its time-coefficient from 'not yet,' or 'not quite yet,' to 'just gone' or 'gone,' as it passes by," (James, 1890, p. 630). As he proceeds, James emphasizes more and more the incessantly "fading" or "passing" nature of "instants of consciousness" (James, 1890, p. 630), which he describes in a footnote (paraphrasing a certain M. Guyau) as "*trainees lumineuses* in the mind like the passage of shooting stars. Each image is in a more fading phase, according as its original was more remote. This group of images gives duration, the mere time-form, the 'bed' of time," (James, 1890, pp. 632–633).

So it seems that, by the end of the chapter, James has settled on a view that sees a succession of momentary representations, progressively fading into the past and renewed from the future, grounding a structure of experience *as if* it were extended over time. This structure remains invariant "like the rainbow on the waterfall," as events pass through consciousness (James, 1890, p. 630). According to such a view, the duration block of the specious present results from the character of the contents of representations, which at any moment simultaneously represent things as being past, just-past, imminent, etc., and which are *not themselves extended*. These are the seeds of Husserl's more fleshed out view of retentionalism, as we shall see later. A further possible interpretation is that James was introducing what we would now call a "hybrid" model, with experience literally extended over an interval in addition to having representational content grounding the experience of "passing" time. Ultimately, though, it is not necessary to settle the question of what James himself was thinking; we need only see that the specious present as presented in *The Principles of Psychology* is amenable to several different understandings and is not tied up with any particular classic model.

It is useful, however, to identify a "minimal specious present"—one that any model seeking to do justice to the concept should be able to accommodate. This minimal doctrine would consist of the following propositions: (1) the subjective present is not the same as the objective present, if there even is such a thing (hence "specious"), and (2) the subjective present consists of an interval of time. Criterion 2 can be accommodated in multiple ways: retentionalists would say we experience the present *as* extended, while extensionalists would say our experience is itself extended. A hybrid view would say both ways of accommodating this criterion are ultimately correct, but we will come to that in greater detail in the coming chapters.

4.1.2 Objections to the Specious Present

Besides the ambiguity in the original presentation, some critics of the doctrine of the specious present believe the concept itself and the arguments for it are confused. One vocal proponent of such a view is Simon Prosser, who thinks that "every current version" of the specious present ought to be rejected (Prosser, 2017, p. 147).

Prosser has two main objections. The first is that the notion of the specious present is at the very least unclear, and likely incoherent because of its inadvertent commitment to the spurious Cartesian Theater view. Second, Prosser thinks the arguments made in favor of the specious present are unsound. These objections are ultimately unconvincing, as we shall see. The contention of this chapter, agreeing with James, is that the specious present constitutes a fact of temporal experience that must be accommodated and explained by models of subjective time.

Prosser disputes one specific argument in support of the doctrine of the specious present that has been particularly influential. He identifies C.D. Broad's discussion of subjective time as the archetypal exemplar of this type of reasoning. Broad pointed out there is a distinct phenomenology of change in the sense that we do not experience simply *that things have changed,* we seem to experience *change itself.* For example, "to see a second-hand *moving* is a quite different thing from 'seeing' that an hour-hand *has* moved. In the one case we are concerned with something that happens within a single sensible field; in the other we are concerned with a comparison between the contents of two different sensible fields," (Broad, 1923, p. 351). That we have such a phenomenology of change, involving as it does comparison between two times, presents a *prima facie* rationale for the notion that we must experience the present as an interval rather than as discrete instants, which on their own would not be sufficient for an experience of change.

Based on the kinds of sentiments expressed by Broad, James, and others, Prosser believes we can paraphrase the primary *a priori* argument in favor of the specious present as follows:

> [P1] Motion, and indeed change of any other kind, essentially takes time. [P2] An experience whose content consisted in the state of the world at an instant could only have in its content that which could occur within the instant. [C1] It thus could not include anything essentially extended in time. [C2] Therefore, given that motion essentially takes time but can be perceived, the content of an experience cannot be an instant… similar arguments can be given by appealing to anything perceptible but essentially extended in time. (Prosser, 2017, pp. 147–148)

Prosser thinks such arguments are unsound because he believes P2 is false: the experiential content of an instantaneous state can include non-instantaneous content. As Prosser puts it: "they fail because they mistakenly assume that an experience with an instantaneous content can have, in its content, only that which could occur instantaneously. This is simply false, for extended processes nevertheless have instantaneous parts, even if the nature of such parts depends on what occurs at other times. All that can be concluded from such arguments is that the surrounding temporal context must be detected by the system that produces the experience," (Prosser, 2017, p. 148).

Quite rightly, Prosser does not think the requirement of sensitivity to temporal context rules out the possibility of instantaneous contents that represent change. He believes that instantaneous experiential content can, on its own, include "motion-like" or "change-like" properties that can underpin the experience of change without appeal to an interval. Prosser appeals to the "dynamic snapshot" theory advanced by Valtteri Arstila (Arstila, 2018) to illustrate this possibility. The dynamic snapshot theory is a variation of the traditional but currently unpopular cinematic model, which sees experience as unfolding in a frame-by-frame manner like a film. One of the main problems with the frame-by-frame, or "static snapshot," view is that a series of instants in succession do not seem able to account for the phenomenology of change, because change takes time and instants do not (Prosser, 2017, p. 147). The dynamic snapshot theory hopes to overcome this worry.

Prosser is committed to some form of intentionalism.[6] This is the idea that the phenomenal character of an experience (what it's like) is determined wholly or in part by the representational content of that experience. The notion of content is notoriously fraught, but for now we can accept Prosser's definition that "the content of an experience is what the experience is *of*," (Prosser, 2017, p. 149). According to Prosser, changing and being in motion are "state[s] that something can be in at an instant, even though it can only be in that state by virtue of being in other places at other times. So the motion of an object, including its direction and rate, could be part of the content of an experience, even if the content of that experience concerned only what was the case at one specific time," (ibid.). In other words, the content doesn't need to change for the content to be *of* change. Prosser claims the animus towards the snapshot view is inspired by a failure to understand the latter point. Instead of extending the content of experience, the dynamic snapshot view contends there are vectors, with a rate and direction, encoded at instants, representing all continuous changes, from movement over time to changes in pitch or color.

Prosser appeals to the so-called waterfall illusion[7] (Wohlgemuth, 1911) and similar motion after-effects[8] to justify the dynamic snapshot theory empirically. In these sorts of illusions, viewing non-moving visual scenes (e.g., a static picture) after some kind of sustained movement results in the non-moving visual scene appearing to move. Prosser says that, in these cases, "motion is experienced even though relevant aspects of the content of the experience remain constant," (Prosser, 2017, p. 150).

Prosser's appeal to this empirical data is open to two immediate objections. The first is that the way he describes the phenomenology may be inaccurate. There is no

[6] The terminology surrounding representation and intention is complex and inconsistent between authors. Here "intentionalism" refers to claims that the phenomenal character of an experience is either identical to or supervenient on the representational content of that experience. In some corners, this same idea has been termed "representationalism." It is implied by intentionalism that all phenomenally conscious experiences have representational content. A further form of intentionalism, which we can call "partial" or "weak" intentionalism, maintains only that phenomenal character is partly determined by representational content, and leaves open the possibility of phenomenally conscious states with no representational content. All of these are in contrast to anti-intentionalism or anti-representationalism, which claims representational content does not determine phenomenal character and phenomenally conscious mental states needn't have representational content at all.

[7] An early description of this illusion can be found in Part 2 of Aristotle's *De Insomniis* (On Dreams) in the *Parva Naturalia* (1957).

[8] Cf. Exner, 1875 and Thorson et al., 1969. An example can be found here: http://www.michaelbach.de/ot/mot-adapt/index.html.

mistaking the so-called "motion" of a motion after-effect for motion under normal circumstances over time. Even to call it "motion" is a little misleading—it could perhaps more accurately be understood as "warping of the visual field". Whatever we call the phenomenal character of the motion aftereffects, we do not generally confuse such experiences with experiences of actual motion, probably because they are simply not experiences of motion. Ultimately, the characterization of the phenomenology of motion after-effects is contentious, and while the illusion seems to me quite unlike veridical motion, that is not to say everyone will have the same intuition (indeed, Prosser does not). The next chapter will analyze the waterfall illusion in much more detail in order to make the case that motion aftereffect phenomenology does not support the dynamic snapshot theory, but for now it is enough to say it is certainly not unequivocal evidence for the plausibility of the dynamic snapshot theory.

Secondly, it is not clear that the content of the experience in the motion aftereffect case is not changing over time after all, contrary to Prosser's contention. If the subsequent static image seems to become increasingly warped or move towards the subject as a result of exposure to previous motion, then there *is* a change in the experiential content, even if the physical image does not change. Changes in experiential content may not require a change in an external object to occur, but there is nonetheless a change in the content, and the illusion gives us no further reason for thinking that this change has been represented at an instant. Again, the waterfall illusion will be considered in much greater detail in the next chapter, but these initial considerations may at least make us reconsider how easy it is to dispense with the motivations for the specious present.

Unlike Arstila (2018), Prosser is ultimately not much interested in whether he is telling the right story about the waterfall illusion. What is important to him is that "the very possibility of such an explanation illustrates the possibility that computational processes in the visual system could yield the content that the object *is moving* (with velocity *v*) at a specific time," (Prosser, 2017, p. 150). For what it's worth, this is correct. The dynamic snapshot theory is a possibility. But, and this is something Prosser does not deny, the specious present is also a possibility. What we are interested in, though, is the plausibility of the view as a description of our *actual* experience. If there are independent reasons beyond *a priori* arguments for establishing the specious present as an actuality, then, practically speaking, we should pay more attention to it than the snapshot theory. So far it seems to me there is little to recommend snapshot theories in general beyond their mere logical possibility.

That Prosser admits the dynamic snapshot view "cannot explain the experience of discontinuous changes" does not help the cause either, especially when the specious present arose in the first place as a simple way to accommodate all kinds of change in an intuitive, phenomenologically consistent way. Discontinuous change (e.g., an immediately perceived change from blue to green) is a problem for the dynamic snapshot theory because the vector encoding at instants would not be relevant when there is no rate or direction of change at any instant. Rather than reject the dynamic snapshot theory, Prosser thinks it is instead more likely that the brain employs various different systems in the detection of different kinds of changes.

In fact, the doctrine of the specious present has empirical evidence behind it as well. The literature on postdictive effects suggests there is something like an interval of time experienced as present where percepts might all be considered to be "now", but the temporal order is not yet fixed. By physiological necessity, there must also be a "window of temporal integration" that allows information from various modalities to synchronize, and the instant-by-instant view seems unable to accommodate this as well (see Vroomen & Keetels, 2010, for an overview of empirical literature regarding this "window").

One thing Prosser makes sure to acknowledge is that "perception has a finite resolution," and this applies to time as well as space (Prosser, 2017, p. 148). Just as there is a minimal level of detail we can detect through vision, so there is a minimal period of time within which we can't tell the order of things. Now it may be that for many who uphold the specious present doctrine, especially those less troubled with the philosophical implications, the idea of a "minimal temporal resolution" is simply what they mean by the "specious present," and it is the duration of this minimal resolution that the experiments have uncovered. This kind of interval, though, is importantly different from the specious present as it does not necessarily satisfy criterion 2 of the minimal definition outlined earlier. The specious present requires that the experienced present occupy an interval of time, but it could be there is a minimal temporal resolution that is not experienced *as* an interval, but rather all of its content is as of a single moment, which would be contrary to the specious present view. For the minimal temporal resolution to be a specious present, we must experience that finest interval as an interval, which is not necessarily implied by the notion of a minimal temporal resolution, however likely of a consequence it may seem.

For Prosser, misplaced faith in the doctrine of the specious present is a symptom of an even greater conceptual confusion that Dan Dennett originally brought to light (Dennett, 1993, pp. 101–111). This confusion is the often tacit but supposedly widespread commitment to the "Cartesian Theater" model of perception, which involves a privileged seat for the subject, who, like an audience, watches as things occur on the perceptual stage. Dennett originally disputed that such a conception accurately captures conscious experiences, which he thought actually do not occur to the subject at determinate times but only ever as a procession of competing "drafts" with no endpoint or completed scene for a subject to "view." As Dennett explains, in this Multiple Drafts Model, "all varieties of perception—indeed, all varieties of thought or mental activity—are accomplished in the brain by parallel, multitrack processes of interpretation and elaboration of sensory inputs. Information entering the nervous system is under continuous 'editorial revision'," (Dennett, 1993, p. 112).

However, in the case of time things are not the same as with garden-variety percepts. Prosser seems to make the mistake of thinking time is like one of these many other inputs we detect that the wrongheaded Cartesians believe play out on the stage—in other words, for Prosser, the metaphor of the Cartesian Theater applies to time just as it does with other experiences. Indeed, he seems to equate the experience of time with the experience of change. For example, on Prosser's view, the detection of "transients" ("rapid changes in luminance or colour of the retinal image") constitutes the detection of time in a way that explains our experience of discontinuous change (Prosser, 2017, p. 151). But time is not like other things that can be detected by sense organs. It is pervasive—rather than being a player *on* the stage for us to view, it is the scaffolding that makes both the Cartesian Theater and the competing multiple drafts model possible. While there doesn't seem to be any obvious reason why the specious present shouldn't be compatible with experience ultimately being constituted by multiple drafts, it *does* seem like any kind of cinematic or snapshot model invites the Cartesian Theater metaphor—on such views, after all, we experience a series of instants that pass like frames on the screen of a cinema.

Another positive feature of the specious present is that it remains neutral on the status of representations. Whereas Prosser talks extensively of "content," an anti-representationalist could still maintain there is a specious present without thinking about experiences as having representational content. What we think of as "now" might really be an interval of time. Does this mean the contents or the vehicle of representation, or both, are extended over an interval, or that the interval itself is the content of representation? We need not commit to a strong position on representations of time while accepting the specious present, and, indeed, as we shall see later, it is perhaps better that we don't.

4.2 The Cinematic Model

In contrast to specious present views, the cinematic model contends we are directly aware of events as they occur on a momentary basis, as if we were watching a film. In the standard formulation, "our streams of consciousness are composed of continuous successions of these momentary states of consciousness," (Dainton, 2023). The actual succession of momentary states is supposed to account for the experience of succession. The cinematic model has its virtues, with simplicity foremost among them. We have a series of instantaneous experiences mirroring a series of instants in the external world, just as we might expect pre-theoretically. The setup is intuitive and the analogy with films or flipbooks is easy to understand. It represents a default, naïve, or common-sense understanding of the structure of subjective time. As Ian Phillips puts it, taking naïveté seriously means that "the stream of consciousness inherits the temporal structure of the events that are its contents. You 'take in' the temporal structure of the events you are witnessing in witnessing them,"

(Phillips, 2014a, p. 139).[9] Because of this simplicity, in the absence of countervailing considerations, we might be inclined to hold onto something like the cinematic model.

Problems with the cinematic model emerge quickly upon closer inspection, however. Probably the most discussed defect is that the model fails to explain the experience of succession. As Kant,[10] James,[11] Husserl,[12] and many others[13] have pointed out, an actual succession is *insufficient* to produce an experience of succession. As James Mill put it, "One idea, upon this supposition [that of a mere, actual succession], would follow another. But that would be all. Each of our successive states of consciousness, the moment it ceased, would be gone forever. Each of those momentary states would be our whole being," (Mill, 1869, p. 369). Besides being insufficient, neither is actual succession *necessary* for the experience of succession. Daniel Dennett somewhat famously illustrated this by disentangling representational content from the representational vehicle of that content. We can take "representational vehicle" to refer to "the sub-personal, functional, or neurobiological realizers of experience," which bear the content of experience, understood as whatever the experience is of, e.g., particular objects and properties (Kiverstein & Arstila, 2013, p. 446). As we have seen, Dennett's claim is that "the representing by the brain of *A before B* does not have to be accomplished by: first: a representing of A, followed by: a representing of B. The phrase 'B after A' is an example of a (spoken) vehicle that represents A as being before B, and the brain can avail itself of the same freedom of temporal placement," (Dennett, 1993, p. 149).[14] The logical possibilities for how the brain generates temporal phenomenology are therefore not constrained by the actual temporal order of worldly events.

[9] Phillips also attends to illusions: on a naïve view, there is still matching between experience and "the *apparent* temporal structure of the objects of experience," which, he thinks, is all the naïve view requires. He also rightly clarifies that not all cinematic views are naïve (one more sophisticated view will be discussed here later), and not all naïve views need be cinematic, so long as there is a "*direct* link between the temporal properties of perception, and its temporal content [emphasis added]," (Phillips, 2014a, p. 145).

[10] Kant (1787, p. B46).

[11] James expresses this idea particularly eloquently: "…between the mind's own changes *being* successive and *knowing their own succession*, lies as broad a chasm as between the object and subject of any case of cognition in the world. *A succession of feelings, in and of itself, is not a feeling of succession. And since, to our successive feelings, a feeling of their own succession is added, that must be treated as an additional fact requiring its own special elucidation*," (James, 1890, pp. 628–629).

[12] Husserl (1917/1964).

[13] Including Thomas Reid (1785), Sir William Hamilton (1861), Shadworth Hodgson (1878), Wilhelm Volkmann (1884), Rudolf Hermann Lotze (1887), Maurice Merleau-Ponty (1945), among others.

[14] The problem of succession does not have to be stated in the language of representation, although it is most apparent when it is. Sequences or series abound in the environment (take any series of events in time or space as examples—a row of trees or a playlist of songs), but a mere succession of elements, as far as we can tell, does not carry with it any subjective experience of one element coming after another.

Although Dennett and his predecessors have opened up the logical possibility that the temporal order of events in the world is radically and systematically at odds with the order of our experiences, it must be noted that too radical a divergence, too often, would seem to be implausible as an adaptive solution to life's problems. However, our concern is primarily with the order of experiential events on short timescales, namely, timescales near the minimum required for the experience of successive events. On these timescales, empirical observations have indeed recorded some very counterintuitive anomalies in the experience of temporal order. Unfortunately, the classic cinematic model has trouble accounting for such phenomena because it maintains that experiences are independent of temporal context, with each experiential event just matching a corresponding worldly event and not further shaped by events around it.

Postdictive effects represent a particularly prominent and thorny class of apparent temporal mix-ups for cinematic views to deal with. Examples of postdictive effects include the flash-lag illusion (MacKay, 1958) and apparent motion (or "phi" effect, cf. Kolers, 1972). In one version of the flash-lag illusion, a red square moves from the left to the right side of a black screen. At exactly the time the red square reaches the middle of the screen, a green square flashes directly underneath the red square. However, rather than perceive one square lined up precisely on top of the other as it is presented in reality, subjects report experiencing the red square ahead of the green one, which should not be possible if experiences match the events in the world in a 1:1 way, as on the cinematic model. Instead, it seems the further motion of the red square that takes place *after* the flash of the green square appears to influence the experience of prior events, namely, the perceived position of the red square at the time of the flash itself.

The example of apparent motion illustrates a similar problem for the cinematic model. In one type of apparent motion, color phi (previously mentioned in Chap. 2), a subject is shown a blue dot in the upper left corner of a screen, then a blank screen, and then a red dot at the bottom right-hand corner of the screen. Rather than two dots, subjects report seeing a blue dot *move* across the screen to the bottom left corner and change to red in the middle of its path across the screen. Here is a case in which the subsequent presentation of the red dot seems to affect the perception of what occurred previously, i.e., the movement of the blue dot and change of color on its way to its destination in the bottom right corner. However, this phenomenon only takes place within a certain timeframe, as Kiverstein and Arstila describe:

> If the temporal gap between *A* and *B* [the two stimuli; in the above case, blue and red dots] is too short (less than 50 milliseconds) we are likely to perceive the stimuli as simultaneous. If the interval is too long—400 milliseconds or longer—the illusion is weak and *A* and *B* will usually be perceived as two distinct stimuli. When the temporal gap is between these limits (it works best when the interval is between 50-200 milliseconds), *A* and *B* are fused into a single stimulus *C* and we experience the illusory or apparent motion of *C* from the location at which *A* is initially presented to the location of *B* (Kiverstein & Arstila, 2013, p. 446).

A much more commonplace and less surprising example of apparent motion occurs when a flipbook is flipped through or film is played. In this case, the existence of subsequent frames or pictures appears to affect how we view previous ones, e.g., whether we view them as in continuous motion or static. If a frame or page is seen on its own, which is to say, without an event on either side of it (i.e., a very low or zero frame-rate), the image is seen as static. However, if the image occurs in a context where subsequent, changing frames occur within a short enough temporal remove of the original frame, the image is seen to move.

While a naïve cinematic model has no way to easily account for these kinds of phenomena, cinephiles can simply deny that postdictive effects are what they seem to be at first glance. The problem for the cinephiles is that the experience should be set as soon as the requisite processing takes place, which means they would predict, to take the case of color phi, an experience of a blue dot in one corner and a subsequent experience of a red dot in another place. However, rather than admit future events or anticipations thereof are influencing the character of the experiential present "moment," proponents of the cinematic view might claim that we actually *don't* experience a blue dot changing to a red dot and moving continuously from one corner to the other. Instead, we experience it just as they predict and then later *revise* our memory of the event, leaving no trace of the original experience to report to experimenters or reflect upon. This process would all have to take place extremely quickly for us never to be aware of the original experience at all or not be able to "catch" it and report it. While the jury is still out about the cognitive mechanisms responsible for postdictive effects, the "memory wiping" response looks strained when compared to those of the other models, which have no problem accommodating for these phenomena.

Alternatively, the cinephile can claim that the final experience is the result of subpersonal processes. This leads to another problem, however, which is that during the intervening time between the appearance of the first dot and second in actual time, we should still be experiencing something. If that "something" is not the continuous movement of the dot and its changing color halfway (and it should not be, on the cinematic view, if future events are irrelevant to present experiences), then why do we only have an experience as of the continuous movement of the dot and its changing color halfway? The subpersonal process response leaves us in the unfortunate position of "waiting" for the appearance of the next dot before we have any experience. To avoid such an unpalatable outcome, it seems the cinematic model again has to invoke a form of memory wiping to explain why we do not have the experience it suggests we should have in the color phi case. Apart from saving the view, it is unclear why such ad-hoc memory wiping would take place in cases where postdictive effects are observed and not in normal cases.

Besides the difficulty accounting for the phenomenology of succession and postdictive effects, the cinematic model also has little to recommend it in accounting for many of the other phenomenal features we associate with subjective time. A succession of instantaneous experiences is difficult to reconcile with the phenomenology of continuity, for example. What is it that allows these instants to feel like they are streaming rather than like a disjointed series, when the latter is the actual phenomenological situation according to the cinematic model?

Cinematic theorists have a number of possible answers to this question. A particularly implausible response is that all of these phenomena are the results of inferences made on the basis of "conceptual knowledge combined with sensible clues to past and future to be found here and now at the scene of the time," (Arstila & Lloyd, 2014b, p. 77). Unfortunately, such a strategy conflates two very different kinds of subjective experiences. Inferring from memory is a very different kind of experience from the kind of vivid phenomenology that the models of subjective time generally try to capture. As in Broad's example of the hands of the clock, it is not merely that we surmise the second hand is continuously moving in the same way we surmise the hour hand must be, i.e., on the basis of the hand being seen in two different places at two different times. Rather, we have a further, vivid experience of continuity that calls out for explanation. Inference does not appear to be the right mechanism for producing the appropriate phenomenal character of our experiences.

A more parsimonious response for the defender of the cinematic view is to appeal to temporal resolution. Discrete instants of experience seem to us as if they are continuous with each other just because their succession occurs too quickly for us to notice, in much the same way that we are unable to see individual pixels on a screen or frames of a movie.[15] If low temporal resolution is responsible for experiences of continuity, though, then we should expect that in some conditions (e.g., neurological disorders) all of our experience might slow down to a speed where its fundamentally discrete character becomes readily apparent. As of yet, conditions like these remain elusive.[16] That being said, the temporal resolution reply remains a live possibility.

Another way of answering how continuity arises is through direct appeal to memory. By invoking memory, adherents of the cinematic model can claim that, while experiences themselves might be discrete, they are bound together in memory, which furnishes us with a continuous narrative of events for us to consciously reflect upon. Traces of events stored in memory are also helpful in establishing a sense of trajectory, in that current events can be compared with previous ones to establish a direction of "flow." The iterative fading of events into the past is supposed to be explained in this way as well. This "memory view" does however seem to stand in contrast with our experience, which, in the present moment, certainly doesn't *seem* to be an artifact of memory access, but instead a rather more direct connection with our sensory apparatus. The case of those who are severely memory-impaired is also instructive here—though unable to recall old memories and form new ones, nonetheless amnesiacs experience the passage of time as others do.[17]

---

[15] The temporal resolution response dates back at least to Thomas Reid (1785) and Dugald Stewart (1792).

[16] Akinetopsia, also known as "visual motion blindness," (discussed in more detail in the next chapter) is an extremely rare condition that is sometimes reported to have this disjointed, strobe-like phenomenal character, but is actually a visual processing deficit typically resulting from lesions to the occipital lobe. Akinetopsia does not affect other modalities or the subject's global experience of time—subjects still report experiencing that time has stopped *for a certain length of time*, indicating that temporal experience continues as normal even though the visual image might be static. See Zeki (1991) for a review of this unusual phenomenon.

[17] See Corkin, 2013 for an accessible study of amnesia with specific attention to the famous case of Henry Molaison (H. M.), who was the author's patient.

While we can distinguish between kinds of memory, e.g., episodic, procedural, and working memory, among other proposed divisions and taxonomies, it seems clear at least the first two of these are poor candidates for the job asked of them by the "memory view" permutation of the cinematic model. Working memory is most promising in this regard but also the least well understood. Working memory is supposed to be active only over short timescales and does play a role informing our present behavior. A worry with this idea, however, is that it seems the memory view collapses into a model wherein nearly all, or maybe even all, of subjective experience *is* just working memory. While this allows for the possibility that the memories comprising most of our present experiential content can potentially shift around in their order, it also means a departure from the naïve view that gave the cinematic model its appeal.

An additional, more conceptual concern with the cinematic view, at least as it has been formulated here, is whether it even makes sense to speak of "instantaneous" or "momentary" experiences. Physical constraints on neural processing speed set a bottom limit on the duration of clock time that will have elapsed between discrete experiences, precluding a matching of successive moments of objective time (assuming there are such things) with subjective time. Although some sort of matching could occur at a temporal remove from objective time, this leaves out a potentially infinite number of "instants" of objective time that would have occurred between discrete experiences. The notion of an experience arising "at an instant" likewise presents difficulties if we think that processes involving activity in many areas of the brain result in conscious experiences. After all, it takes some non-zero amount of time for any neuron to transmit electrochemical signals, meaning the many interacting processes responsible for experience must take some amount of real time for their influence to propagate.

It is wise not to see the processes as completed and packaged as discrete moments of experience. To think there is some "endpoint" of processing that occurs at a particular instant (and in succession) falls prey to the fallacy of the Cartesian Theater, suggesting some "homunculus" or further subject is viewing the final experiences, and the explanatory regress that follows. Indeed, the metaphor of the cinema that gives this model its name suggests this very problem.[18] A related worry is that positing a succession of experiences to explain phenomena like continuity is really akin to positing a further, internal flipbook to explain our experience of flipbooks. Ultimately, the cinematic model lacks a decent explanation for how it can result in the expected temporal phenomenology and appears to create more problems than it solves.

[18] While not everyone agrees that Dennett's Cartesian Theater represents a problematic understanding of the way the mind works, the prospect of the infinite regress that comes from positing a subject "watching" successive frames of experience as if at the cinema should at least be some cause for concern (see Dennett, 1993, p. 107).

4.3 Extensionalism

Extensionalist models claim our experience exhibits an actual extension over time, and this fact explains our temporal phenomenology. As Dainton puts it, on these views, "our episodes of experiencing are themselves temporally extended, and thus able to incorporate change and persistence in a quite straightforward way," (Dainton, 2023). Rather than a succession of instants, as in the cinematic model, in extensionalist models our experience "extend[s] a short distance through ordinary clock-time, in just the way [it] seem[s] to," (Dainton, 2017). For extensionalists, we apprehend an extended present and this explains why the present seems to have duration (i.e., a specious present), as James suggested it does. This is in contrast to retentionalist views in which we do not apprehend an extended present but apprehend the present *as* extended.

Extensionalism can be construed in a direct realist way if it is seen as maintaining that we directly apprehend an interval of time without representing that interval; that is, the experience itself is extended and this gives the experience the character of being extended. Indeed, one of the major motivations for extensionalism is that we needn't appeal to representational content to explain temporal phenomenology. This feature lends the model more credibility for those that consider the notion of representational content nebulous or explanatorily impotent with respect to phenomenology.

4.3.1 The Discrete Block Model

The Discrete Block Model is an extensionalist model that simply sees experience as "a succession of specious presents laid end-to-end in the manner of a line of building blocks, or bricks in a wall," (Dainton, 2017). The experience of a sequence of sounds would be something like the series A-B, C-D, E-F…, where each pair represents a specious present and each specious present is arranged end to end. Because the sections are up against each other but do not overlap, the model is called discrete. Notable advocates of the Discrete Block Model include Sprigge (Sprigge, 1983) and possibly Whitehead (Whitehead, 1929). In order to account for the direct perception of change and succession (i.e., the thesis of "phenomeno-temporal realism") the extended temporal intervals (the specious presents) are said to contain more than one event that is experienced in succession or as changing. This co-consciousness of events over the duration of the specious present is meant to allow for the phenomenology of succession, continuity, and change. This model therefore has the advantage of simplicity, accommodation of the specious present, and an explanation of the phenomenology that occurs within individual duration blocks.

However, in the Discrete Block Model, the phenomenology is *not* accounted for over multiple specious presents. In essence, this model runs into the same problem that the cinematic model does in connecting its momentary, static snapshots. A succession of specious presents, in and of themselves, like a succession of moments, is not sufficient to ground an experience of succession, continuity, change, etc. In the Discrete Block Model, we end up with "a *succession of experiences*, but there is no *experience of succession*" as Dainton puts it (Dainton, 2017). In the case of continuity, the situation is the same: on the one hand, the gapless structure of the Discrete Block Model is enough to give us a real continuity, but not *an experience of continuity*. If PT-realism is true, then the Discrete Block Model insufficiently accounts for the basic phenomenological features of temporal experience.

4.3.2 The Overlap Model

The Overlap Model is a kind of extensionalist model designed to overcome the problems of the Discrete Block Model. It has been advanced by Bertrand Russell (Russell, 1915), John Foster (Foster, 1979), and Barry Dainton (Dainton, 2006), among others. The problem of accounting for continuity across the transitions between specious presents is dealt with here by invoking further specious presents (duration blocks) that overlap each other. Rather than only a discrete series of specious presents A–B, C–D, E–F, etc., the Overlap Model posits further specious presents, B–C, D–E, etc., overlapping the others. Importantly, this does not involve a redundant tokening of the parts (e.g., two D phases) but rather the overlapping specious presents share common parts with the specious present they overlap. There are no *additional* experiences beyond the sequence posited in the original Discrete Block Model, rather there is an additional *experiential relationship* (Dainton, 2023). In other words, in the newly posited overlapping specious present, the set of experiences that is experienced as co-conscious is expanded to include those in the overlapping duration blocks (B–C, D–E, etc.). As Dainton puts it, "it is now being supposed that D and E are connected by the relationship of diachronic co-consciousness, whereas it was previously held that only C–D and E–F were so related," (Dainton, 2017). However, not *all* experiences are diachronically co-conscious on such a model, but only those that share in an overlapping specious present (e.g., A and F are not co-conscious, for that would be phenomenologically inaccurate). As it stands however, this model is too simple. The overlapping specious presents can include *parts* of any given experience too, so in a more complex Overlap Model there could be a dense overlay of specious presents accounting for varying overlapping intervals of potentially differing lengths across and within different modalities (limited by the brain's physical constraints).

Barry Dainton (2006) has proposed a version of the Overlap Model that makes some important further qualifications that stem from insights about the nature of experience in general. Dainton devotes special attention to the notion of co-consciousness, which extensionalists maintain holds both synchronically and diachronically. For Dainton, attempts to describe this feature of experience in terms of anything more basic will fail because co-consciousness of any kind is primitive. Diachronic co-consciousness, i.e., the "togetherness" of experiential objects over time, is therefore a feature that other phenomena can be explained in terms of, but will not be explicable in terms of other more fundamental phenomena. On his view, co-consciousness should not be understood as the unification of "experiential objects and contents [...] under a single act of awareness" in the way that prior overlap models have alleged (Dainton, 2023). Dainton thus denies the "awareness-content structure" commonly adopted by so many of his predecessors, whether extensionalist or non-extensionalist. On Dainton's view, there is simply no need for both awareness and content when, simply in virtue of being experiential, phenomena are experienced. As Dainton puts it, "contents are themselves intrinsically conscious, and hence—in a manner of speaking—they are self-revealing or self-intimating. That is, phenomenal contents become conscious simply by coming into existence," (Dainton, 2000, p. 57). This is a powerful insight that we will return to in Chap. 6.

Dainton also emphasizes the intrinsic "directional dynamism" that results in "phenomenal flow," experienced as a kind of continual renewal of the experience. This feature is absent from other extensionalist models, but is, in my view anyway, a real phenomenological explanandum. In terms of the Overlap Model, this manifests as the individual elements, say musical notes, exhibiting future-directionality and flow through time. The view is thus not an overlap of static elements in diachronic unity with each other, but inherently dynamic in the way that it is constituted. This addition redresses an oversight in the phenomenological description that earlier Overlap Models had subscribed to.

There are several advantages to this type of extensional view. According to the Overlap Model, change and succession are experienced directly and vividly, as they seem to us and in accordance with realism about these phenomena. The Overlap Model takes it that change and succession are apprehended just as well as other phenomenological features like shape and color, and can be considered "immediately apprehended" as opposed to "inferred." Furthermore, there is phenomenological continuity within and across specious presents. This is because the source of these phenomenological features is just the relationship between normal experiences, and therefore no higher-level inference is required. A phenomenologically accurate picture of which experiences would be diachronically co-conscious and which would not be (i.e. those not adjacent) also emerges from this picture. At the same time, the overlaps explain how phenomenological features like continuity can persist *across* and not only *within* specious presents. Indeed, this focus on continuity is supposed to be a particular strength, which is not accommodated at such a core level in cinematic models, retentionalism, the discrete block, or other variations of extensionalism.

However, there are also drawbacks to the Overlap Model. We might think, as Shaun Gallagher and Galen Strawson do, that the model does not have phenomenological support, because the overlapping specious presents are undetectable (Gallagher, 2003; Strawson, 2009, Ch. 5). Inherent in the model is the assumption that any phenomenological indication of overlap must be absent because if it were not, that would mean phenomenal continuity hadn't been achieved. However, the fact of phenomenal continuity is the very thing the Overlap Model was enlisted to explain. There is therefore no way of determining, on the grounds of experience, if such a model is correct, and there is no support to be found in experience for the model. It appears then that the Overlap Model was not generated directly out of a consideration of temporal phenomenology, but rather on *a priori* grounds as an answer to what must be the case for continuity to hold given that we have temporally extended experiences. There is something apparently *ad hoc* about this formulation. Although the proponent of the Overlap Model thinks that phenomenological continuity is itself evidence for the view, this could equally be evidence for other non-extensionalist models, such as retentionalism—making any such "evidence" neutral between the positions.

It might also be that the Overlap Model is unnecessary. As Ian Phillips has pointed out, variations in the content of experiences alone (e.g., changes in the trajectory of a bird overhead) can be sufficient to differentiate duration blocks (Phillips, 2010). We needn't appeal to further, subjectively undetectable relationships between parts of experiences to explain this phenomenology. However, we still need to answer why the immediate perception of change and persistence only occurs in short timeframes. If the overlap theorist maintains that the diachronic co-consciousness relationship itself is restricted, then it seems the overlap view can meet this challenge (Dainton, 2023). Compared to the Discrete Block Model, then, the restriction to short timescales and the straightforward accounting of the phenomenology of continuity ultimately leaves the Overlap Model as the more compelling extensionalist model.

4.3.3 Reflections on Extensionalism

Further objections can be leveled against extensionalism and its particular construal of the temporally extended specious present. One problem, dating back to Meinong, is that of the unity of experiences within the specious present (Meinong, 1899). The problem is how the many parts of the specious present, which are spread out in time, could be unified as they seem to be in experience. Without this unification, we might expect experience would seem disjoint, but it is not. In essence then, we are left with the very problem the specious present was meant to solve in the first place. Attempts to solve this problem often look like exercises in regress. For instance, one may think that the unifier is the "single episode of awareness" within which the experiences are co-conscious, but the parts of that episode of awareness likewise occur successively in time, so the problem of *their* unity now rears its head. A further appeal to "a higher-order act awareness" unifying the episodes of first-order awareness runs into the same problem. However, if the original awareness of the temporal parts of experience is not extended but momentary, there is no regress. Such a view seems to endorse the previously mentioned Principle of Simultaneous Awareness (PSA), which states that all the various temporal aspects of an experience are present simultaneously in consciousness (i.e., within a momentary awareness) so that experience can have its dynamic quality. The problem with this line of thinking, however, is that we are essentially discarding the original temporal extension that is distinctive of extensionalist models.

The extensionalist can avoid the above outcome by rejecting the PSA altogether. They can simply say that co-consciousness of temporal phases *needn't be limited* to momentary awareness but really can be extended in time. Indeed, that is the main conceit of the extensional specious present: the episodes within it are diachronically co-conscious, but not necessarily synchronically co-conscious. The contention is that there is no good reason for thinking phenomenal unity cannot be achieved over a temporal interval just as easily as it might be achieved synchronically at a given moment. We can therefore reject the supposition that "synchronic phenomenal unity—the sort of unity which obtains among simultaneous momentary contents—is the only mode of phenomenal unity," (Dainton, 2017). We might wonder how exactly the extensionalist thinks this works, but this is perhaps no more of a problem than explaining how synchronic phenomenal unity works—both require some heretofore inexplicable feats of integration on the part of the brain. Nonetheless, if the phenomenology seems to suggest a temporally extended specious present, as James initially thought, this is a fact in need of explanation and not explaining away. The extensionalist therefore rejects the privileging of moments as the only temporal units capable of unity. In doing so, extensionalism accords with a view of distributed brain processes as *taking time* and not finishing their job by generating individual experiences on a moment-to-moment basis. The latter runs the risk of falling into the trap of the Cartesian Theater mentioned earlier in the discussion of the cinematic model, which we might think is an unrealistic vision of the way the brain works. Extensionalism, for its part, avoids such worries.

A further objection to extensionalism is that the notion of a temporally extended specious present gets the phenomenology wrong on the level of experiences that occur within the interval of the specious present (Pelczar, 2010a, 2010b). The idea is that, because the extended specious present is supposed to account for the phenomenology of succession, change, and continuity, on the specious present view, none of these aspects exist for momentary experiences *within* the specious present itself. Moments are therefore not dynamic but static if we endorse the extensional specious present. However, as Pelczar points out, there are no such discernible static experiences at any timescale. Our experience appears to have the phenomenology of succession, change, continuity—in short, dynamism—all the way down, and extensionalism does not account for this fact in a satisfactory way, in contrast to retentionalist models. However, the extensionalists can avail themselves of an easy response, namely that, indeed, such static moments *are* impossible and *don't* exist in our experience. They are phenomenologically inaccessible because we only ever experience the unity of the specious present and not the component pieces within it in isolation.[19] Another option is to just see Pelczar as begging the question and hold that, in fact, experiences are *not* essentially dynamic. We may not experience them as such because of the way they are unified in the specious present, but this does not mean they are essentially dynamic, only that they are always experienced that way

[19] This response echoes Husserl's contention, in reference to his retentionalist view, that the individual *parts* of the structure of temporal experience can never be isolated—we only have experiences as a temporal unity (Husserl, 1917/1964).

by us because of the large-scale temporal structure. Another, albeit counterintuitive, response is that punctate experiences, static or not, are in fact impossible because, as Dainton puts it, "consciousness has the structure of *atomless gunk,*" and this gunk is "infinitely divisible, but not possessing point-like parts" (Dainton, 2017). We might think something about the physical actuality of the neural substrate of consciousness rules out this scenario, but in any case, we find that extensionalism has several plausible ways of handling the objection that it gets the character of micro-experiences wrong.

To my mind, the idea that consciousness is not atomistic, but essentially processive and dynamic, is rather intuitive. For some, the biological substrate also has this processive character, which removes the atomism of the physiology as a point of objection—the brain, we are increasingly finding, looks less and less like a serial processing digital computer or machine the more we discover about it.[20]

4.4 Retentionalism

According to retentional models, "our experience of change and succession occurs within episodes of consciousness which themselves lack temporal extension, but whose contents present (or represent) temporally extended intervals and phenomena," (Dainton, 2017). The elements of the temporal structure in a retentional model exhibit intentionality in Brentano's sense: *directedness* towards the future, past, or present (see Brentano, 1874, p. 92). Consequently, unlike under the cinematic model, in the retentional model the contents of consciousness "*appear* to [but don't] possess a brief temporal depth [emphasis added]," (Dainton, 2017). However, like the cinematic model, the retentionalist does not see experiences as temporally extended, but momentary. Alternatively, a retentionalist may see experiences as having some temporal extension, but this extension would be completely irrelevant to our phenomenology, in contrast to extensionalism. The retentionalist endorses the Principle of Simultaneous Awareness and insists that the various temporal aspects of an experience are present simultaneously at each moment. The apparent temporal extension of the present, i.e., the specious present, results from the temporal character of intentions or representations, which carry content *about* the later, earlier, and simultaneous events, at each moment.

While the idea of retentionalism might predate the twentieth century (as we saw, a view like this can be seen in James), retentional models really took off following Husserl, who coined the term "retention." Like James, Husserl realized the experience of duration is not entailed by "a succession of isolated, punctual, conscious states," (Gallagher & Zahavi, 2008, p. 75). *Now* for us must be more than a monadic time-slice with no relation to what came before or what will come after because

[20] This point is hotly debated but see Chirimuuta (2024) on this question in neuroscience, and the collection *Everything Flows*, edited by Nicholson and Dupré (2018), for some reasons to think processual thinking is more appropriate for biological systems in general.

such a momentary, fleeting instant could never be experienced as part of the rich temporal flow we find ourselves in.

Husserl took his investigations beyond James, however, and focused particularly on the internal structure of the experienced present itself (i.e., what James called the specious present). Husserl held that merely experiencing an interval is not enough to explain how we experience time as passing. As Gallagher and Zahavi put it, the experience of duration must essentially involve "more than that which is given right now—it must be co-conscious of that which has just been, and that which is just about to occur," (ibid.). Husserl therefore proposes a threefold internal structure of present experience including what he calls "protentions," "retentions," and "primal impressions," (Husserl, 1917).

Our immediate perception includes within it information about what came before (retentions), information about what will come next (protentions), as well as "primal impressions," which are "directed toward the now-phase of the object," (Zahavi, 2003, p. 83). All of these elements are directed at different temporal phases—namely, past, present, and future—and therefore exhibit intentionality, at least in the minimal sense of "direction towards an object," (Brentano, 1874, p. 95). It is only the temporal structure of the whole unit, with forward-looking and backward-looking edges, that imbues experiences with the impression of time's passage. A static moment, without the other elements, could not in itself constitute an experience, for experiences necessarily occur over time, as a passage. Thus it must be that, as Evan Thompson puts it, "the threefold structure is invariant and present synchronically as a unified whole throughout the course of experience," (Thompson, 2010, p. 319).

In Husserl's scheme, protentions are constantly slipping into the present, the present is constantly fading away into retention, and retentions to further back retentions, along with potentially infinitely more fine-grained movements. In this way the present is constantly becoming the past and the past is constantly becoming the further past, and so on. Dainton expresses a simple characterization of retentions consistent with most retentional views: "retentions [are] a special form of past-directed mental representations that are triggered in an automatic (and involuntary) manner after each momentary phase of experiencing," (Dainton, 2017). Unlike retentions, which have definite intentional content specified by the just-past temporal phase, protentions are "unfilled or indeterminate" as they concern yet-to-occur events, but still "involve a sense of anticipation," (Thompson, 2010, p. 319).

It is important to reiterate, for the retentionalist, all the phases of the temporal structure of experience are present *at once* (that is, they exhibit co-consciousness). This is not to say they result in an experience of simultaneity, however, but rather constitute an experience exhibiting different temporal aspects at the same time (e.g., the bow and stern of the specious present to which James alluded). Despite intentionality giving us an experience *as of* an extended specious present, "all variants of the [retentional] approach have this much in common: they compress the experiencing of change into momentary (or near-momentary) phases of consciousness," (Dainton, 2017). While extensionalists see some kind of symmetry between the extension of experience and extension over clock time, retentionalists tend to see a radical divergence in this regard: for retentionalists an extended experience can occur in a moment of clock time.

4.4.1 *Reflections on Retentionalism*

Extensionalism is clearly the more intuitive option for those that think we can directly and immediately perceive succession, continuity, and the other phenomenological desiderata outlined earlier. Extensionalists accommodate these phenomena by allowing that our experiences fundamentally consist of units of duration of real time. This alignment of experience and the world seems to be a rather natural way to accommodate temporally extended phenomenology. For Dainton, the intuitive appeal of extensionalism speaks quite strongly in favor of that view, whereas "By contrast, the Retentional view can seem almost perverse: why think that our apparently immediate experience of change and persistence is in fact packaged into momentary (or extremely brief) slices of experience? Since it is by no means obvious how this can be, Retentionalism is *not* the obvious way to go," (Dainton, 2017). It is therefore important for retentionalists that they provide reasons for why their view is superior to extensionalism or for why extensionalism is implausible.

Retentionalists face an explanatory burden in accounting for how a momentary experience can exhibit the phenomenal character of an extended experience. As discussed above, this is most often done through appeal to intentions/representations, but this appeal in itself might be considered a point of weakness if we think the notion of representation is overly vague or explanatorily impotent with respect to the specifics of the phenomenal character for which it is responsible. However, depending on preexisting metaphysical commitments, sticking with moments can also be considered an advantage for the retentionalist. For subscribers to presentism, which is the claim that only the present really exists, a non-extended view of experience like retentionalism is practically a requirement. At the very least, an interval view like extensionalism, where consciousness is spread out over objective time, is not so obviously compatible with a metaphysical commitment to presentism.

Perhaps a more compelling advantage of retentionalism is the strong commonsense intuition in favor of the Principle of Simultaneous Awareness. It seems that, for us to be conscious of anything, including temporal phenomena like succession, all the contents of consciousness must be present simultaneously at each moment—otherwise, they would be inaccessible to consciousness. To take an example, an experience of a continuous, flowing piece of music in the present must include, at any given moment, some kind of information about which sounds came before and which sounds are expected to come after, in addition to the knowledge *that* sounds have come before and are expected to come after. Without that information, experiences would be disconnected and unrelated. The retentionalist view maintains that multiple intentions (i.e., retentions, primal impressions, protentions, retentions of retentions, etc.) are held in mind at the same time, and these all constitute our present experience, giving it the distinctive temporal character which it otherwise could not have. While it may be that co-consciousness across an interval, as the extensionalists assert, is possible, retentionalists need only appeal to synchronic unity, i.e., phenomenal unity at one moment. Their position is thus simpler in at least one sense of not having a further, arguably more counterintuitive kind of co-consciousness to explain.[21]

Rick Grush believes another major advantage of retentionalist models is that they can account for postdictive effects like the flash-lag effect better than extensionalism (Grush, 2005). Recall that these counterintuitive situations involve later stimuli affecting prior perception, which *prima facie* should be impossible according to a naïve cinematic view. The problem for extensionalism is that the percept only occurs once and is not subject to change over the course of the temporal interval. Our experience now, on such a view, should not be influenceable by future events. On the other hand, retentionalism allows for aspects of future events to influence the character of the present experience. For retentionalists, a dense succession of retentions-impressions-protentions comprises our experience, so if one of these elements changes with incoming information, it can affect the whole of the present experience. For instance, a red circle presented in an opposite corner of a screen from a previously presented green circle may alter retentions in such a way that the whole present experience has a different character than it otherwise would. More specifically, expectations, which might be fundamental both to protentions and retentions, might shape both of these in light of incoming information, altering the resulting experience as a whole. The plethora of intentions involved in the retentionalist view thus allows for a malleable phenomenology.

One extensionalist response to the problem of postdictive effects is that the lag in perceptual processing time allows these effects to occur (e.g., Dainton, 2008). In this way, later stimuli can affect processing such that the resultant experience changes as conflicting, new information enters the system, but only if it arrives before the processing of a "whole" experience is complete. Such an appeal has the advantage of explaining why such effects only occur within extremely short timescales (usually sub 200 ms), where retentionalists seem to have a problem accounting for why postdictive effects don't occur over much longer timescales.

A final problem for the retentionalist is the problem of surplus content. Dainton explains the essence of the problem: "If retentions have the same force and vivacity as ordinary (presently occurring, first-order) experience, as they do in the non-modal variant, there is a risk of flooding consciousness with enormous quantities of surplus content — content which we have no reason to suppose exists," (Dainton, 2017). All that content carried by retentions and the other intentional entities just doesn't seem to figure in experience in the same kind of way as present experience;

[21] I don't mean here to comment on the merits of the PSA, although I do hope to have at least illustrated the intuition behind it. It should be noted that some extensionalists view the PSA as an unjustified dogma (see Stern, 1897/2005, p. 313), which is potentially question-begging against their view. Here I just mean to point out that there appears to be an intuition in favor of the PSA, even if it turns out to be false, and that the principle has been used to motivate retentionalism. Thank you to anonymous reviewer for pointing out the extensionalist perspective on this issue.

rather, it seems, as James noticed, more "faded" in some sense. The retentionalist response to this problem is to endorse a "modal" form of retentionalism, whereby retentions and their kin are *not* like ordinary experiences. Indeed, historically this has been the most prominent form of retentionalism, with Husserl speaking of retentional contents being experienced as "more" or "less" past, and not as present.

4.5 Comparison

It is now possible to compare the classic models and assess their strengths and weaknesses. Owing to the numerous objections previously considered, an unsophisticated version of the cinematic model would appear to be the least convincing of the classic models. The cinematic view's most commendable feature, simplicity, is just not enough to make up for its other shortcomings in explanatory power and consistency with phenomenology.

Extensionalism on the other hand maintains simplicity in its appeal to real temporal features of experience without needing to rely on intentional content. The present just feels extended because our experience is itself extended. While co-consciousness within the span of the specious present can account for certain aspects of temporal phenomenology within that interval, extensionalism runs into problems accounting for things like continuity over and above the duration of one specious present. How the specious presents are connected to resemble something like a holistic experience becomes a major problem for extensionalism's viability. Essentially, the problems of succession and continuity encountered by the cinematic view are just moved one step up, so that now we talk of problems of the integration of specious presents. One way around this, which cinephiles can also avail themselves of, is to simply bite the bullet and accept that experience is fundamentally "chunky" and discontinuous, but we simply don't notice, much as we don't notice visual saccades. However, this seems to disregard an initial phenomenological consideration that calls out for explanation.

Retentionalism, though potentially more extravagant than extensionalism in its commitments to representations and its complicated "nesting" tripartite structure of temporal experience, offers a solution to the problems the other models face. Besides more easily accommodating postdictive effects, retentionalists have recourse to intentions—retentions, protentions, and primal impressions—to accommodate the specious present, doing justice to the Principle of Simultaneous Awareness by appeal to synchronic unity. However, it also seems like retentionalism may go too far in divorcing phenomenology from the constraints of objective time, which we might think should still have *something* to do with subjective experience. Relatedly, it is somewhat problematic that the intentions comprising the invariant and persistent temporal structure of experience are often understood to be momentary or durationless. Beyond this, an integration problem remains for retentionalism, namely the problem of explaining how phenomenal unity is achieved without regress to more and ever-broader representations. The extensionalist, in contrast, has at least already headed off the problem by endorsing a specious present consisting of a single diachronically unified experiential unit.

A further challenge for the retentionalist, though not necessarily a problem, is the question of how retentions, protentions, and primal impressions are physically implemented and how they can furnish us with the rich phenomenal character they are supposed to. Husserl's own initial thoughts were that retentions were a kind of memory (primary memory), distinct from other kinds of memory (secondary memory)(Husserl, 1917),[22] similar to James' early thoughts on the subject. The difference between primary and secondary memory, at least for James, is that whereas objects of secondary memory are dredged up and brought into consciousness, objects of primary memory are "never cut off from consciousness" (James, 1890, pp. 646–647). A claim like this sounds empirical in that there should be identifiable brain processes underpinning this "primary memory" view of the temporal structuring of consciousness. What these brain processes might be remains an open question. Extensionalists, by not positing similar entities, do not face a similar empirical challenge. Rather, an extensionalist can say that whatever brain process is responsible for experience comes pre-loaded with a temporal aspect just in virtue of itself being extended over time.

The most fundamental point of disagreement between retentionalists and extensionalists seems to concern the role of intentions or representations in each respective scheme. There is no necessary symmetry between the properties of the contents of a representation and the properties of the vehicle of a representation. Words, for example, are representational vehicles that do not usually exhibit the same properties as their content (e.g., the word "red" is not always itself red, though it can be). Retentionalists exploit this fact in claiming that contents exhibit temporal properties, which give us a familiar temporal phenomenology consisting of things like extension, persistence, and succession, while the vehicles themselves do not exhibit such properties.

Extensionalists, on the other hand, see no asymmetry between the properties of vehicles and contents—the properties of the vehicles are the properties of the content. Wolfgang Köhler, as interpreted by Kiverstein and Arstila, may be an example of someone who rejects the content-vehicle distinction with his contention that "the neural correlates of temporal experience treat time as its own representation," (Kiverstein & Arstila, 2013, p. 447). In the color phi case, this means that "the neural events (n_1 … n_n) correlated with our experience of the red flash must occur before the neural events (m_1 … m_2) correlated with our experience of the green flash," (ibid.). Just because content and vehicle properties are in principle separable does not necessarily mean that is the most plausible situation.

Further, an extensionalist might claim there is no distinction between vehicles and content because representations are not involved in the generation of temporal phenomenology. What an experience is *of* is the content of that experience (a

[22] Husserl's views on this subject were not stable, and his conception of retentions changed throughout his writings, none of which did he take to be definitive.

particular percept, say), but it is only the *way that content occurs to us* that is temporal, e.g., over an interval of clock time. The content itself does not consist of represented temporal properties. So, while the retentionalist would explain the experience of succession as a succession of contents, the extensionalist explains the experience of succession as a succession of experiences.

More important perhaps than this difference with respect to content/vehicle properties, is that retentionalists and extensionalists seem to differ with respect to the relationship between phenomenal and physical properties. Retentionalists, again, see no necessary connection between an actual, physical succession (of the representational vehicle, say) and an experienced succession—we may have one without the other. Extensionalists, in contrast, at the very least, see a close connection or even mirroring between the physical temporal properties and phenomenal properties. The experience feels extended because it *is* extended, and this corresponds to an extension in real time of brain processes, which presumably correspond to some degree with the extension in time of the physical processes responsible for incoming sensory input (although some divergence between brain processes and sensed physical occurrences must be possible to account for illusions). Barry Dainton, for example, in his version of extensionalism, appeals to a processing delay to account for illusions (Dainton, 2008). Because discrepancies between clock time and phenomenal time are more the norm than the exception, with subjective time dilation and contraction occurring in a variety of conditions, the extensionalist has a strong incentive to avoid a precise matching with external temporal properties, if not with temporal properties of correlated brain processes.

The distinction between physical and phenomenal properties is not the same thing as what Kiverstein and Arstila call an "experience-reality distinction" (Kiverstein & Arstila, 2013, p. 445). It is quite clear that there are many cases of temporal illusions and temporal variability, so it would be a difficult position to maintain that in all and every case temporal experience matches reality. Nonetheless, Ian Phillips (2014a, 2014b), in maintaining a form of extensionalism that endorses the view that the temporal properties we experience are fully determined by the temporal properties of the experience itself (a kind of inheritance view), is committed to a view like this, at least insofar as the experiential properties match the real properties of the experience. Even in illusory cases, according to Phillips, the temporal properties of the experience are inherited, and this is possible because on a pure form of extensionalism we must be committed to the idea that temporal properties of the entire diachronic span of the specious present are still determinative and can be inherited by the experience just as well (Phillips, 2014a). While this may be, it still may seem inexplicable on such a view why, e.g., postdictive effects should have the *particular* illusory character they do in virtue of the properties of the particular temporally subsequent events that occur within the relevant specious present. While it may not be impossible to give such an explanation, it would appear the appeal is at least not as straightforward as that available to those who maintain a vehicle-content distinction.

It is a somewhat easier case to make that temporal experience matches temporal properties of brain processes,[23] which themselves might not line up neatly with the temporal properties of events in the world (e.g., a brain process might take a long time, resulting in a longer experience, even though the event that was experienced occurred very quickly). There are indeed some instances where temporal illusions are easily explained without even detaching physical properties from phenomenal properties at all, and it might be tempting to think of these as a model. For example, hearing thunder after seeing lightning is a non-veridical temporal illusion, but results from basic physics (light travels faster than sound). Perhaps similarly, there are neural processing time considerations which can result in experiential differences, e.g., touching one's nose and toes at the same time does not result in feeling the two sensations simultaneously because of differences in processing time (Kiverstein & Arstila, 2013, p. 445).

In further defense of extensionalism, we might suspect that temporal properties are profoundly different than run-of-the-mill perceptual properties, making the content/vehicle distinction much less relevant in this case. Time, to borrow Rodney Brooks' well-known phrase, is most plausibly "its own best model" on this view (Brooks, 1991). Against Dennett's emphasis on the divergence between content and vehicle properties, Ian Phillips points out the uniqueness of time—that it is essentially different from properties like greenness:

> Experiences which represent green trees do not themselves have to be green. Nor of course is Emily Dickenson's reference to Spring—"This whole experiment of green"—itself green! Yet it is hard to see how this bears on the case in point. Time is *uniquely* common to experience and its objects. Consequently, there is no clear analogy at all between the representation of green in experience (or poetry) and the representation of simultaneity or duration in experience. Experiences do not have colour properties; they do have temporal properties (Phillips, 2010, pp. 34–35).

If the temporal properties of experience are there to be exploited, why wouldn't they be? Why propose temporal properties of content when the vehicles have temporal properties that can already give their contents some temporal structure? There is an apparent synergy here, which retentionalism seems to ignore in favor of a more baroque approach rooted in the potentially spurious notion that time is a property like any other, or simply "more content." It is one thing to say there is no necessary connection between the properties of a vehicle and that of its contents, but to say that *in actuality* there is no connection is a greater leap.[24]

There are further reasons for thinking time might be special, and quite disanalogous to properties like color. Kant determined that time must be a precondition of any experience whatsoever. Our concept of time is *a priori* in that it is not derivative of our experience of the world. We can see this when we consider that "neither coexistence nor succession would ever come within our perception, if the representation of time were not presupposed as underlying them *a priori*," (Kant, 1787, p.

---

[23] Georg Northoff (2016, 2023) for example grounds a form of extensionalism in the physiological oscillation patterns of the brain, i.e., brainwaves.

[24] For much more on this issue see Chap. 6.

B46). In other words, we cannot deduce coexistence and succession from worldly experience, for that worldly experience cannot be had without already being structured according to principles like coexistence and succession.

The necessity of time for the possibility of all other experiences is apparent when we consider that, "in regard to appearances in general one cannot remove time, though one can very well take the appearances away from time. In it alone is all actuality of appearances possible," (Kant, 1787, p. A31). According to Kant, we should think of time "not as an object [of a representation] but as the *way of representing* myself as object [emphasis added]," (Kant, 1787, p. B54). Time is thus dissimilar to things like tones and colors, because temporality is common to all of consciousness, and consciousness is inconceivable without it. The notion that time should be treated like other contents, or relations between contents, and that it is not rather a structural feature of consciousness, is questionable and may be considered a potential downside of the retentional view—namely that it does not treat time as adequately "special."

Retentionalism also faces a potential challenge accounting for the finite interval of the present. For extensionalists, the interval that we take to be the present is the specious present, and it is of some more or less determinate length. Retentionalism, taken on its own, does not specify a cut off or definable interval, but seems like it allows in principle for retentions and protentions ad infinitum in either direction. Furthermore, the notion of discrete, instantaneous representations seems dubious when it comes to any kind of realistic instantaneous physical implementation. Brain processes occur over time, so it seems as though intentions generated by those brain processes also ought to have some non-zero duration and should properly be considered processes themselves and not states.

4.6 Conclusion

To sum up, we can see that extensionalism has problems accounting for continuity, while retentionalism has problems coherently accounting for the specious present. Appealing to aspects of both views can give us something closer to a model that does justice to the way we construct our experiential present. Experiences themselves may have explanatorily relevant temporal properties, as extensionalists maintain, but the way we represent experiential content may also matter. The combination of these properties could better account for our temporal phenomenology in all its richness, as we shall see when the advantages of drawing on both kinds of explanation are discussed in Chap. 6. However, before that, the next chapter will address why recent proposals to rehabilitate the snapshot/cinematic model are misguided and why we should instead continue to (1) favor the specious present view, and (2) look beyond localized cognitive mechanisms and instead look towards systemic ones when thinking about how the phenomenology of the specious present is constructed.

References

Andersen, H. (2014). The development of the 'specious present' and James's views on temporal experience. In V. Arstila & D. Lloyd (Eds.), *Subjective time: The philosophy, psychology, and neuroscience of temporality* (pp. 25–42). MIT Press.

Arstila, V. (2018). Temporal experiences without the specious present. *Australasian Journal of Philosophy, 96*(2), 287–302.

Aristotle. (1957). *On the Soul, Parva Naturalia, On Breath.* W. S. Hett, Trans. Loeb Classical Library, 288. Harvard University Press.

Arstila, V., & Lloyd, D. (2014a). Subjective time: From past to future. In V. Arstila & D. Lloyd (Eds.), *Subjective time: The philosophy, psychology, and neuroscience of temporality* (pp. 309–322). MIT Press.

Arstila, V., & Lloyd, D. (2014b). Contemporary philosophies of lived time. In V. Arstila & D. Lloyd (Eds.), *Subjective time: The philosophy, psychology, and neuroscience of temporality* (pp. 75–81). MIT Press.

Berkovich-Ohana, A., & Glicksohn, J. (2014). The consciousness state space (CSS): A unifying model for consciousness and self. *Frontiers in Psychology, 5,* 341. https://doi.org/10.3389/fpsyg.2014.00341

Brentano, F. (1874/2015). *Psychology from an empirical standpoint*. Routledge Classics.

Broad, C. D. (1923). *Scientific thought: A philosophical analysis of some of its fundamental concepts*. Kegan Paul.

Brooks, R. A. (1991). Intelligence without representation. *Artificial Intelligence, 47*, 139–159. https://doi.org/10.1016/0004-3702(91)90053-M

Chirimuuta, M. (2024). *The brain abstracted: Simplification in the history and philosophy of neuroscience*. MIT Press.

Corkin, S. (2013). *Permanent present tense: The unforgettable life of the amnesic patient, H. M.* Basic Books.

Dainton, B. (2000). *Stream of consciousness: Unity and continuity in conscious experience.* Routledge.

Dainton, B. (2006). *Stream of consciousness: Unity and continuity in conscious experience* (2nd ed.). Routledge.

Dainton, B. (2008). Sensing change. *Philosophical Issues, 18*, 362–384.

Dainton, B. (2011). Time, passage and immediate experience. In C. Callender (Ed.), *Oxford handbook of philosophy of time* (pp. 382–419). Oxford University Press.

Dainton, B. (2014). The phenomenal continuum. In V. Arstila & D. Lloyd (Eds.), *Subjective time: The philosophy, psychology, and neuroscience of temporality* (pp. 101–137). MIT Press.

Dainton, B. (2017). Temporal consciousness. In E. N. Zalta (Ed.), *Stanford Encyclopedia of Philosophy* (Fall 2017 Edition), https://plato.stanford.edu/archives/fall2017/entries/consciousness-temporal/

Dainton, B. (2023). "Temporal consciousness". In E. N. Zalta and U. Nodelman (Eds.), *The Stanford Encyclopedia of Philosophy* (Spring 2023 Edition). https://plato.stanford.edu/archives/spr2023/entries/consciousness-temporal/.

Dennett, D. (1993). *Consciousness explained.* Penguin.

Exner, S. (1875). Experimentalle Untersuchung an der einfachsten psychischen Processe. *Pflügers Archiv, 11*, 403–425.

Foster, J. (1979). In self-defence. In G. MacDonald (Ed.), *Perception and identity: Essays presented to A. J. Ayer with his replies to them* (pp. 161–185). Macmillan.

Frost, D., & Pöppel, E. (1976). Different programming modes of human saccadic eye movements as a function of stimulus eccentricity: Indications of a functional subdivision in the visual field. *Biological Cybernetics, 23*(1), 39–48.

Gallagher, S. (2003). Sync-Ing the stream of experience: Time-consciousness in Broad, Husserl, and Dainton. *Psyche, 9* (10). Available online at Retrieved February 23, 2022, from https://journalpsyche.org/files/0xaabb.pdf

Gallagher, S., & Zahavi, D. (2008). *The phenomenological mind*. Routledge.
Grush, R. (2005). Internal models and the construction of time: Generalizing from state estimation to trajectory estimation to address temporal features of perception, including temporal illusions. *Journal of Neural Engineering, 2*(3), S209–S218.
Hamilton, W. (1861). *Lectures on metaphysics and logic*. Hansebooks.
Hodgson, S. H. (1878). *The philosophy of reflection* (1st ed.). Longmans.
Husserl, E. (1917/1964). In M. Heidegger (Ed.), *On the phenomenology of the consciousness of internal time* (1893–1917) (J.S. Churchill. Trans.). Indiana University Press.
James, W. (1890/1980). *The principles of psychology*. Henry Holt.
Kant, I. (1787/2007). *Critique of pure reason* (N. K. Smith, Trans.). Palgrave Macmillan.
Kent, L., & Wittmann, M. (2021). Time consciousness science and its theories. *Neuroscience of Consciousness, 7*(2), niab011. https://doi.org/10.1093/niab011
Kiverstein, J., & Arstila, V. (2013). Time in mind. In H. Dyke & A. Bardon (Eds.), *A companion to the philosophy of time* (pp. 444–469). Wiley. https://doi.org/10.1002/9781118522097.ch26
Kolers, P. A. (1972). *Aspects of motion perception*. Pergamon Press.
Le Poidevin, R. (2015). The experience and perception of time. In *The Stanford Encyclopedia of Philosophy* (Summer 2019 Edition). Edward N. Zalta (Ed.). https://plato.stanford.edu/archives/sum2019/entries/time-experience/
Lloyd, D. (2012). Neural correlates of temporality: Default mode variability and temporal awareness. *Consciousness and Cognition, 21*, 695–703.
Lockwood, M. (2007). *The labyrinth of time: Introducing the universe*. Oxford University Press.
Lotze, R. H. (1887). *Metaphysic in three books: Ontology, cosmology, and psychology* (2nd ed.). Clarendon Press.
MacKay, D. M. (1958). Perceptual stability of a stroboscopically lit visual field containing self-luminous objects. *Nature, 181*, 507–508.
Meinong, A. (1899). On objects of higher-order and their relationship to internal perception. In S. Kalsi (Ed.), *Alexius Meinong on objects of higher order and Husserl's phenomenology* (pp. 137–208). Martinus Nijhoff.
Merleau-Ponty, M. (1945). *Phenomenology of perception* (Colin Smith. Trans.). Routledge.
Mill, J. (1869). *An Analysis of the Phenomena of the Human Mind,* J. S. Mill's Edition. Longmans, Green, and Dyer.
Nagel, T. (1974). What is it like to be a bat? *Philosophical Review, 83*(4), 435–450.
Nicholson, D. J., & Dupré, J. (Eds.). (2018). *Everything flows: Towards a Processual philosophy of biology*. Oxford University Press.
Northoff, G. (2016). Slow cortical potentials and 'inner time consciousness'—A neuro-phenomenal hypothesis about the 'width of present'. *International Journal of Psychophysiology, 103*, 174–184.
Northoff, G. (2023). *Neurowaves: Brain, time, and consciousness*. McGill-Queen's University Press.
Pelczar, M. (2010a). Presentism, eternalism and phenomenal change. *Synthese, 176*(2), 275–290.
Pelczar, M. (2010b). Must an appearance of succession involve a succession of appearances. *Philosophy and Phenomenological Research, 81*(1), 49–63.
Phillips, I. (2010). Perceiving temporal properties. *European Journal of Philosophy, 18*(2), 176–202.
Phillips, I. (2014a). The temporal structure of experience. In V. Arstila & D. Lloyd (Eds.), *Subjective time: The philosophy, psychology, and neuroscience of temporality* (pp. 139–158). MIT Press.
Phillips, I. (2014b). Experience of and in time. *Philosophy Compass, 9*(2), 131–144.
Pöppel, E., & Bao, Y. (2014). Temporal windows as a Bridge from objective to subjective time. In V. Arstila & D. Lloyd (Eds.), *Subjective time: The philosophy, psychology, and neuroscience of temporality* (pp. 241–262). MIT Press.
Prosser, S. (2017). Rethinking the specious present. In I. Phillips (Ed.), *The Routledge handbook of philosophy of temporal experience*. Routledge.
Reid, T. (1785/2002). In D. Brookes and K. Haakonssen (Eds.), *Essays on the intellectual powers of man*. Edinburgh University Press.

Russell, B. (1915). On the experience of time. *The Monist, 25*, 212–233. https://doi.org/10.5840/monist191525217
Sprigge, T. L. S. (1983). *The vindication of absolute idealism*. Edinburgh University Press.
Stern, L. W. (1897/2005). Mental presence-time (Nicolas de Warren, Trans.). In B. Hopkins and S. Crowell (Eds.), *The new yearbook for phenomenology and phenomenological philosophy* (Vol. 5, pp. 310–351). Noesis Press.
Stewart, D. (1792). *Elements of the philosophy of the human mind*. Gale Ecco.
Strawson, G. (2009). *Selves: An essay in revisionary metaphysics*. Clarendon Press.
Thompson, E. (2010). *Mind in life: Biology, phenomenology, and the sciences of mind*. Harvard University Press.
Thorson, J., Lange, D. G., & Biederman-Thorson, M. (1969). Objective measure of the dynamics of a visual movement illusion. *Science, 164*, 1087–1088.
Volkmann, W. F. (1884). *Lehrbuch der Psychologie Vom Standpunkte des Realismus: Des Grundrisses der Psychologie* (3rd ed.). BiblioBazaar.
Vroomen, J., & Keetels, M. (2010). Perception of intersensory synchrony: A tutorial review. *Attention, Perception, & Psychophysics, 72*, 871–884. https://doi.org/10.3758/APP.72.4.871
White, P. A. (2017). The three-second "subjective present": A critical review and a new proposal. *Psychological Bulletin, 143*(7), 735–756. https://doi.org/10.1037/bul0000104
Whitehead, A. N. (1929). *Process and reality: An essay in cosmology: Gifford lectures delivered in the University of Edinburgh during the session 1927–28* (Corr. ed.). Free Press.
Wittmann, M. (2014). Embodied time: The experience of time, the body, and the self. In V. Arstila & D. Lloyd (Eds.), *Subjective time: The philosophy, psychology, and neuroscience of temporality* (pp. 507–524). MIT Press.
Wohlgemuth, A. (1911). *On the after-effect of seen movement* (Vol. 1). Cambridge University Press.
Zahavi, D. (2003). *Husserl's phenomenology*. Stanford University Press.
Zeki, S. (1991). Cerebral Akinetopsia (visual motion blindness): A review. *Brain, 114*, 811–824. https://doi.org/10.1093/brain/114.2.811

Chapter 5
On Dynamic Snapshots and Localized Mechanisms

This chapter[1] extends the defense of the specious present begun in the previous chapter by challenging a recent proposal that would seek to overturn that doctrine. That recent proposal is known as the "dynamic snapshot theory" (or DST) and is meant to upgrade the classic cinematic model to make it more plausible as an alternative to the more popular specious present views, namely retentionalism and extensionalism. However, the dynamic snapshot view is flawed for multiple reasons. The present chapter will focus in particular on how the dynamic snapshot view's purported justification fails and actually supports the specious present view instead. My point in this chapter is not to directly dispute cinematic views such as those of Reid (1785) or Le Poidevin (2007), as I think what has been said previously should have already raised sufficient doubts about that kind of approach. Instead, I am targeting the dynamic snapshot theory in particular because it is, in my view, a more compelling contemporary alternative to specious present views and also because more important general lessons can be drawn from thinking about the arguments employed in its defense.

The dynamic snapshot view holds that we have snapshot-like experiences that encode information about change, motion, succession, etc., at an instant. The dynamic snapshot view therefore rejects the orthodox "specious present" doctrine, which claims the experienced present must be extended in time rather than instantaneous, as phenomena like change, persistence, continuity, and succession require intervals of time. As mentioned previously, motion aftereffects (MAEs) and in

[1] This chapter is based on a previously published paper (McKenna, 2021), but has been substantially reworked here to improve clarity. I therefore invite anyone working in the field to refer to this chapter and not the older paper, which I now consider deprecated.

C. A. McKenna, *Constructing the Present*, Studies in Brain and Mind 24,
https://doi.org/10.1007/978-3-031-89571-5_5

particular the waterfall illusion have been recruited[2] in support of the dynamic snapshot view of subjective time (e.g., Prosser, 2016, 2017; Arstila, 2018). These illusions involve the apparent motion of a static object following a subject's prolonged exposure to moving stimuli. The contention here is that motion aftereffects like the waterfall illusion have in fact been misappropriated and do not provide evidence for the conclusions of the dynamic snapshot theorists.

My argument against the dynamic snapshot theorists' use of MAEs has two strands. First, I will argue that the phenomenology of motion aftereffects should compel us to recognise that subjective temporal properties encoded at an instant are *not* sufficient for a subjective experience of motion, contrary to the position of the dynamic snapshot theorists, who maintain they are sufficient for the phenomenology of pure motion in the illusory case.[3] That this is so will become apparent when the phenomenology of the waterfall illusion is appropriately interpreted. Second, I will argue that MAEs like the waterfall illusion should be seen as illusions of *motion*, and therefore a result of visual processing, rather than temporal illusions *per se*. The rare condition of akinetopsia (motion blindness) helps to demonstrate that the experience of visual motion has no particularly special connection with distinctly temporal phenomenology, so drawing analogies between their mechanisms is misguided. Ultimately, the connection between subjective time and motion aftereffects is too strained for advocates of the dynamic snapshot view to effectively make a case for generalizing to temporal phenomenology.

Underlying the misappropriation of MAEs is a methodological error regarding the kind of evidence that constitutes appropriate fodder for analyses of time-consciousness. The problem is that theories seeking to account for temporal phenomenology are not the kind that can be supported by evidence from any one particular sensory modality or cognitive function, such as visual motion.[4] Proponents of the dynamic snapshot theory miss their target by focusing on a visual motion illusion. As a general constraint on models of temporal phenomenology and explanations of the cognitive mechanisms underlying such models, we should see

[2] From Prosser (2017, p. 149): "The dynamic snapshot theory gains some plausibility from various empirical sources. Perhaps one part of the intuition that snapshot experiences can only be static is connected with the idea that for there to be an experience of change, the content of one's experience must itself change over time. But this appears to be false; there are many examples of motion illusions in which motion is experienced despite the fact that no part of the content of the experience changes (apart from time itself). The best-known example is the waterfall illusion (Wohlgemuth, 1911)…"

[3] This contention goes further than a recent critique of the dynamic snapshot theory by Jack Shardlow, who, while critical of the overall theory, does not take issue with the purported phenomenology of the waterfall illusion (Shardlow, 2019, p. 745).

[4] To be clear, the dynamic snapshot theory does not seek to explain temporal phenomenology merely by appeal to mechanisms underlying visual motion perception. Rather, the theory holds that *something similar* to the mechanisms underlying visual motion could explain all of the temporal phenomenology that philosophers usually think would require an experienced interval (e.g., the feeling of succession), thus obviating the need for an extended experiential present (i.e., a specious present).

temporality as a global precondition for subjective experiences. It is more plausible that temporal structure is an inherent, fundamental feature of the information processing mechanism underlying consciousness (whatever account we give of this), rather than a result of localised processes. This is the most important lesson to come out of this chapter and helps to motivate the turn towards a non-localized information processing story to account for time-consciousness like that which will be developed in greater detail in Chap. 7.

This chapter proceeds by describing, in Sect. 5.1, the phenomenology of the waterfall illusion as traditionally understood, highlighting the difference between illusions of motion and distinctively temporal illusions. Section 5.2 outlines the dynamic snapshot theorists' understanding of the illusion. Section 5.3 gives a different interpretation—one which is more faithful to the phenomenology and which reveals the illusion is not supportive of a snapshot view of any kind. On this interpretation, encoding change-like properties at an instant is insufficient for the phenomenology of motion without the perception of a change in position over time. Section 5.4 demonstrates the disconnect between visual motion and temporal phenomenology by considering the phenomenon of akinetopsia, or motion blindness, and also by delving into the neurophysiological underpinnings of motion perception. Section 5.5 develops the second strand of the argument against the effectiveness of MAEs as evidence for the dynamic snapshot theory. It is argued that MAEs are unrelated to the systems or processes responsible for temporal experience in general. The connection, therefore, between visual motion-like properties of experiential objects and aspects of temporal phenomenology, like succession, is not strong enough to support inferences about the nature of the latter.

5.1 The Waterfall Illusion

An especially vivid early account of the waterfall illusion came from Robert Addams, a Scot who observed the phenomenon at the Falls of Foyers near Loch Ness in 1834. Addams reported that, "having steadfastly looked for a few seconds at a particular part of the cascade, admiring the confluence and descussation of the currents forming the liquid drapery of waters," he then "suddenly directed [his] eyes to the left, to observe the vertical face of the sombre age-worn rocks immediately contiguous to the water-fall," and, upon doing so, "saw the rocky face as if in motion upwards, and with apparent velocity equal to that of the descending water," (Addams, 1834, p. 373). In other words, just a few seconds staring at a moving object (the waterfall), resulted in the perceived motion of a stationary object (the cliff) in the opposite direction and at a speed apparently equal to that of the moving object from which his gaze was diverted.

The illusion does not only apply to waterfalls and cliffs, but also other forms of motion and stationary objects, e.g., spiral motion and dot patterns on a screen.[5] It is telling, however, that experimental researchers and early observers both thought of the waterfall illusion as a *visual* illusion, and not an illusion of *time* in any significant sense. In other words, on the face of it, the illusion has to do with what we *see*, which in this case does not match the state of the world (i.e., the cliff face is not in fact moving). It is not immediately obvious that the illusion concerns how we experience time, although friends of snapshot theories have insisted otherwise (e.g., Robin Le Poidevin, 2007, p. 88; Valterri Arstila, 2018, p. 290; Simon Prosser, 2016, p. 123).

The contrast between MAEs and postdictive effects helps to illustrate the difference between distinctively temporal phenomenology and phenomenology that occurs over time. Visual motion, it will be argued, is an example of the latter. In contrast, phenomena like succession, continuity, and persistence are fundamental aspects of temporal phenomenology, indispensable to subjectivity in a way that motion is not. Postdictive effects are the archetypal illusions of time in the sense that they seem, *prima facie*, to affect the normal experience of succession.

We can focus on the flash-lag illusion[6] as an illustrative example of how postdictive effects concern temporality in a way that MAEs do not.[7] As mentioned previously, in the flash lag illusion, a red square moves from left to right across a screen. When the red square reaches the midpoint, a green square is presented below it at exactly the same horizontal position. However, if the red square *will* continue moving to the right, subjects perceive that the red square is to the right of the green square at the time of the latter's presentation, rather than exactly aligned on top (as it is in fact). If the red square doesn't continue moving right, no illusion is reported.

In the flash-lag illusion, it therefore seems the perceptual system peers into the future before generating the "present" experience. Ostensibly, then, the ordering of perceived events is modulated retrospectively. We have in this case an illusion of succession, such that the perceived event corresponding to the presentation of the green square is perceived to lag that of the moving red square when in fact they are simultaneous. While postdictive illusions appear to be related to the way our brain constructs and orders experiences, MAEs do not concern such things. Rather, MAEs, along with other visual illusions of size, shape, and colour, are temporally normal experiences of things that are not the case.

---

[5] See Macpherson and Baysan (2017) for an excellent online example of a motion aftereffect.

[6] The example described in the main text is from Khoei et al. (2017). See also Bach (2004).

[7] Postdictive effects remain poorly understood and there are many varying interpretations of what is happening in these illusions. Dennett (1993) notably contrasted "Orwellian" (post-hoc memory revision) and "Stalinesque" (modulation prior to experience) accounts of postdictive effects, while himself proposing a "multiple drafts model" of consciousness eschewing a single experiential "end point." For present purposes, I do not mean to weigh in on the correct interpretation of postdictive effects, but only wish to draw out the contrast between illusions standardly conceived as affecting distinctively temporal phenomena (e.g., reordering the normal succession of events, as in the flash-lag illusion) and illusions that are not similarly distinctively temporal in their effects, such as MAEs.

Because visual motion illusions like the waterfall illusion do not concern mistakes related to the temporal structure of experience, and because visual motion is not related to temporal phenomenology in any special sense, we should not see MAEs as temporal illusions at all. There is nothing particularly special about a moving ball as opposed to a stationary ball in the way that we perceive them both lasting through time. We do not perceive a stationary ball as *static in time*, but rather only static in space. Perceived visual properties like colour or motion persist through time just like other perceptual phenomena and have little bearing on the temporality of experience. As Kristie Miller points out, it seems natural to say, "we have the relevant temporal phenomenology even if *nothing* perceptually changes," (Miller, 2019). Indeed, empirical studies do not show that we feel time stops flowing in a sensory deprivation chamber, although this passage feels slower than usual (Wittmann, 2017, p. 125, 127).

In contrast to visual properties of objects, like changes and movements, properties like succession and continuity are deeply and inextricably connected to experience itself. The latter properties are distinctively temporal because these sorts of properties are the *sine qua non* of experience through time—they are what make us think phenomena, whether stationary or moving, are lasting through time in the first place. These considerations should make it clear that visual motion illusions are not *distinctively* temporal illusions in the same way as illusions of succession. Because of this, it is unlikely that they will generate special insights into the mechanisms responsible for temporal phenomenology, even if they are independently interesting phenomena deserving of attention and explanation in their own right.

5.2 The Dynamic Snapshot Account of the Waterfall Illusion

For proponents of the dynamic snapshot view, the waterfall illusion is meant to demonstrate the plausibility of a model of subjective time according to which experience can be analysed as successive instants or moments that exhibit temporal properties like succession, continuity, and persistence. We can recall that, after Addams diverted his gaze from the Falls of Foyers, he saw a static object, the neighbouring cliff face, appear to move. The dynamic snapshot theorist takes it that static moments, like static objects, can exhibit such dynamic qualities as Addams perceived. Our experiences of succession, continuity, persistence, change, and motion are not unlike the illusory motion of the cliff face in this way: what is actually a procession of static snapshots is imbued at each step with qualities that lend a certain non-static appearance to each moment.

Before considering the dynamic snapshot proposal in more detail, it is important to understand where it comes from and why it is appealing. All snapshot views, dynamic or otherwise, maintain the temporal structure of our experience is akin to a sequence of frames in a flipbook, except in this case each frame is an instantaneous experience. According to Barry Dainton's influential formulation, the "classic" snapshot view, which he calls the cinematic model, holds that "our immediate awareness lacks any (or any significant) temporal extension, and the same applies to the contents of which we are directly aware—they are akin to static, motion-free 'snapshots' or 'stills'. Our streams of consciousness are composed of continuous successions of these momentary states of consciousness," (Dainton, 2017). The snapshot analogy is in fact somewhat imperfect, as snapshots do not truly capture instants but rather the interval that the camera shutter remains open (Le Poidevin, 2017, p. 320). Nonetheless, the notion that snapshots do capture "moments" is at least commonplace in the popular imagination.

The perennial popularity of snapshot views[8] is perhaps attributable to the simplicity of atomic theories[9] and the intuitiveness of the analogy with film. If one thinks the mind-independent world proceeds as a series of moments, like movie frames, it seems reasonable to suppose that our experience of the world has a similarly "cinematic" temporal structure.[10] However, to account for the further *experience of* succession, appealing to a mere succession isn't enough, for we might just as easily experience each event as entirely unrelated to the others. Thomas Reid, responding to John Locke's simple snapshot view,[11] therefore claimed experiences of succession arise through the reflective comparison of experiential snapshots via memory (Reid, 1785, p. 271). Reid, in a rare departure from his usual reverence for common sense, thus endorses what Dainton calls phenomeno-temporal antirealism (PT-antirealism). This means that for Reid, succession, change, and duration are not *directly* experienced (Dainton, 2017). Unfortunately, Reid's view doesn't fit well with the way we usually take our experiences of succession to be, as they do not obviously require reflection or memory retrieval to occur. One way to get around this issue is to reject the snapshot view entirely and adopt a specious present view—one in which a window or interval of consciousness can take in successive events together. However, for those enamoured with the simplicity of the snapshot picture, another alternative is to tweak the theory.

---

[8] For contemporary *non-dynamic* snapshot theorists, see Francis Crick and Christoph Koch (2003), Robin Le Poidevin (2007), and Philippe Chuard (2011).

[9] Hoerl (2017, p. 100, n. 20), discussing Chuard (2011), treats "atomism" as another name for snapshot theories, but there is a subtle distinction. Whereas atomism is concerned with the decomposition of experience into primitive units or "atoms," snapshot theories are committed to the latter *and* to experience being structured as a series of instants. One could hold an atomist view that treats the indivisible, fundamental temporal "atoms" of experience as extended in time, rather than instantaneous. See Piper (2019, p. 2), for a concise description and list of atomists. For these reasons, the "atomist" nomenclature, which has been at the center of a debate in the metaphysics of time-consciousness (see, e.g., Lee, 2014), will be avoided here.

[10] Unfortunately, the simple cinematic analogy is fraught with peril. For instance, one non-trivial difference between films and experiences is that, whereas there is someone watching a film in a cinema, in the experiential case it seems the snapshot theorist has to say something like we *are* the film, or else risk an infinite homuncular regress.

[11] As a snapshot theorist living before the invention of snapshots, John Locke thought ideas were like "images in the inside of a lantern, turned round by the heat of a candle" (Locke, 1690, 2.14.9). See Hoerl (2017, p. 94, note 12) for an illuminating description of how Locke's lantern might have worked.

With a view to preserving more common-sense notions of temporal phenomenology and account for our seemingly direct experience of change and succession, the *dynamic* snapshot theory adds that the instantaneous snapshots of experience have dynamic qualities. More precisely, Valterri Arstila, the progenitor[12] of the dynamic snapshot theory, identifies four essential theses of the dynamic snapshot view (Arstila, 2018, p. 291). These are:

(1) realism about temporal phenomenology (PT-realism)
(2) punctuality of phenomenal contents
(3) purity of temporal phenomenology
(4) encapsulated mechanisms

As previously mentioned, phenomeno-temporal realism (PT-realism) is the claim that, "we have immediate experiences of change, motion, and other temporal phenomena, as the majority of philosophers claim," (Arstila, 2018, p. 291). Under PT-realism, there is a phenomenological difference between experiencing the second-hand of a clock changing position and noticing, perhaps via inference from memory, that an hour-hand *has changed positions*, with the latter not immediately part of our present experience in the same way as the former (Broad, 1923, p. 351). The divergence between the dynamic snapshot theory and standard snapshot theories is stark here. As Hoerl (2017) argues, the standard snapshot theorist cannot draw a principled distinction between experiences of the second hand and the hour hand if both kinds of experience are based on memory, *a la* Reid. By endorsing PT-antirealism and denying the phenomenological distinction between experiences of the second hand and hour hand, standard snapshot theorists apparently ignore a basic *explanandum*. In contrast, the dynamic snapshot theorist hopes to avoid this oversight by accounting for the second hand with dynamic snapshots, while holding there is nothing dynamic about the hour hand.

Arstila's second thesis constitutes the core of any snapshot view. This is the claim that experience is ultimately composed of snapshot-like instants. Such a thesis amounts to a denial that experience is extended over time in the way non-snapshot views would have it, e.g., retentionalism and extensionalism, which see the present as consisting of an interval known as the "specious present." Formulated in an intentional way, according to the dynamic snapshot theory, the *contents* of experience are not extended over time. This is unlike retentionalism, where the contents are extended, or extensionalism, where the content/vehicle distinction might collapse but the extension of the experience itself is preserved.

The third thesis (we can call it "the purity thesis") maintains that temporal phenomenology is possible without "an associated phenomenology of things being different at different times" (Arstila, 2018, p. 291). In other words, according to the purity thesis, we only need to experience one instant to have a phenomenology of

---

[12] Arstila (2018) uses the term "dynamic snapshot view," which he calls both "my view" and "my theory". I have opted to use the term "dynamic snapshot theory" or "DST," but I believe these are interchangeable. While Arstila is not the only defender of DST, he coined the term and has defined and defended the view precisely, situating it relative to other snapshot and specious present views.

change, motion, succession, continuity, and whatever else is thought of as belonging to temporal phenomenology. Dynamic snapshot theorists take the waterfall illusion to support this thesis. By challenging the dynamic snapshot theorist's use of the waterfall illusion, this chapter seeks to dispute the purity thesis.

Lastly, the fourth thesis, concerning encapsulated mechanisms, contends that, "temporal phenomenology is brought about by primitive mechanisms, each separate from the other" (ibid.). This means that temporal phenomenology is the result of many different mechanisms, perhaps different ones for each modality, as well as aspects of cognition. Such a position multiplies the complexity of any possible account of subjective time, ruling out explanations that seek to ground temporal phenomenology in fundamental structural features of any single overarching mechanism responsible for consciousness. The waterfall illusion is meant to support the thesis of disparate primitive encapsulated mechanisms by demonstrating that motion phenomenology, as one aspect of temporal phenomenology, is produced by vector encoding occurring in the visual cortex, while other aspects of temporal phenomenology would be the result of similar mechanisms elsewhere. Ultimately it is my position that, of the four theses that define the dynamic snapshot theory, only the first, phenomeno-temporal realism, is true. Before arguing against these theses however, I will first show how proponents of DST have appealed to the phenomenology of the waterfall illusion to justify them.

Advocates of the dynamic snapshot theory use the waterfall illusion as evidence for the claim that features of temporal phenomenology can be experienced in isolation without requiring an extended, experiential specious present. The waterfall illusion supposedly demonstrates this by showing that even a static object, in the absence of a perceived change in position over time, nonetheless can appear to change or move. As Arstila puts it, "...the most significant aspect of this explanation [of the waterfall illusion] is that the experience of motion is explained in a framework where the contents can, subjectively speaking, be confined to an instant," (Arstila, 2018, p. 290). Simon Prosser likewise claims that, "'Moving' is a state that something can be in at an instant," (Prosser, 2017, p. 149) before leaning on the waterfall illusion for empirical support. To put it in the terms of Addam's visit to the Falls of Foyers, for proponents of DST the cliff face would take on an immediate sensation of motion without requiring any time to elapse.

Although the purported phenomenology of "instantaneous motion" does not appear in the descriptions of Addams and others, Arstila nevertheless thinks this is what is happening in such situations based on the cognitive and neural mechanisms thought to be responsible for the illusion. Arstila's idea here is particularly indebted to Robin Le Poidevin's description of the processes underlying MAEs. Le Poidevin (2007, p. 89) identifies two neural mechanisms, one that detects motion "by a change in retinal stimulation" and another that "register[s] the relative position of an object and store[s] it in the short-term memory for comparison with later perceptions of its relative position," resulting in perceived change when these diverge.[13] Le Poidevin believes the first mechanism is more "primitive," does not require any change in position, and gives rise to a sense of the change occurring "now." He thinks, following Gregory (1966), that the first mechanism results in so-called "pure motion" phenomenology in that it does not concern a relation. The second mechanism, meanwhile, "employs short-term memory, takes a series of snapshots of an object's relative positions, and compares them," (Le Poidevin, 2007, p. 89). Le Poidevin extends this analysis to temporal phenomenology generally, such that, for instance, "the conjunction of the very recent memory of C [a musical note] with the perception of E [another musical note] gives rise to an experience of 'pure succession'," (ibid., p. 91). This kind of extrapolation from the case of visual motion is a move that proponents of the dynamic snapshot view will also take up.

Le Poidevin is committed to a more traditional (non-dynamic) snapshot view. Consequently, he maintains the only difference between the illusory case and the case of veridical motion is that, while in the former only the first mechanism is engaged (i.e., we experience motion without a change in position), in the normal case both mechanisms contribute to motion phenomenology. Arstila and Le Poidevin agree that motion and change should be considered aspects of temporal phenomenology. However, whereas Le Poidevin, like Reid, sees the comparative function afforded by short-term memory as crucial, Arstila's fourth thesis rules that out. Indeed, Arstila is explicit that "the [fourth] thesis separates the dynamic snapshot view from Le Poidevin's account, according to which the second mechanism involved in the waterfall illusion is short-term memory and 'the conjunction of the very recent memory of the [first stimulus] with the perception of the [second stimulus] gives rise to an experience of 'pure succession' (Le Poidevin, 2007, p. 91)," (Arstila, 2018, p. 290). For Arstila, unlike Le Poidevin, "both pure motion and pure succession are explained by appealing to a primitive mechanism specific to those experiences. For example, the second mechanism involved in the waterfall illusion is likely to be a second-order motion processing mechanism [not memory] (Mather et al., 1998)," (Arstila, 2018, p. 291). The dynamic snapshot theorist thus appeals to a further higher-order mechanism that is derivative of or parasitic on the first-order primitive mechanism responsible for pure motion and succession. Of course, if the supposed phenomenology of pure motion and pure succession turn out to be nothing of the kind (as I will argue), this move becomes much less compelling.

Like Arstila, Simon Prosser (2016) argues the waterfall illusion supports the dynamic snapshot view in that it supposedly provides an instance of a static object exhibiting motion. Prosser believes one of the reasons philosophers generally don't think experience can be punctate in the way demanded by a snapshot view is because it seems that for a full-blooded temporal phenomenology, including the experience of change, succession, and the like, we must have an experience "including different states at different times" (ibid., p. 123). The thinking here is that the two states must be contained within a single experience for the transition between them to be

[13] Note the similarities between Le Poidevin's description of this second mechanism and Reid's earlier conjecture about the role of memory.

directly perceived in experience (i.e., PT-realism), and it is this thinking that motivates specious present views.

For Prosser, as with Arstila, the waterfall illusion challenges the specious present view by demonstrating that phenomena like motion, change, etc., do not require an extended experience, but only a snapshot. Based on the illusion, Prosser argues vector encoding of perceptual information (i.e., encoding of "both the rate and direction of motion") could be the mechanism responsible for the experience of motion (Prosser, 2016, p. 124; Prosser, 2017, p. 149), which allows for instantaneous states to exhibit properties like change over time.[14] He then generalises this mechanism, like Arstila and Le Poidevin, to offer an account of "all such continuous perceptible changes" of any sort. Prosser (2017, p. 149) explicitly offers the model of "instantaneous vector rate of change" as an alternative for the phenomenology of change to the specious present view, which he views as unnecessary. Notably, however, Prosser does not think such a mechanism can accommodate discontinuous changes like the sudden change of a light from on to off.

Prosser, with his focus on the connection between temporal experience and metaphysical views of time, is more interested in establishing the mere possibility of a mechanism that eschews the specious present. Despite the arguments he presents for the dynamic snapshot view, Prosser is careful not to fully commit to it. That said, Prosser does claim that, "for experiential content there is no logical entailment from the lack of temporal extension of the content to the content containing only what is 'static', or lacking change" (Prosser, 2016, 122). The logical possibility of such a world can be granted. However, it is of crucial importance to an adequate understanding of motion phenomenology, let alone subjective time, that such a world is not *our* world, as we shall see.

5.3 Visual Motion

Until the recent resurgence of snapshot theories, the waterfall illusion derived most of its fame in philosophy from its paradoxical character. As several philosophers have noted, the initial description of the illusion, in which the stationary object is said to exhibit *motion*, is misleading (Blakemore, 1973; Frisby, 1980; Crane, 1988). Tim Crane, for instance, observes that, "[A]lthough the stationary object appears to

[14] See work by Alan Johnston and colleagues supporting a vector-encoding mechanism on the information processing side of visual motion perception (Johnston et al., 1992, 1999). Johnston contends that "Motion is represented at a point and at an instant from a calculation over a spatial region and an extended period of time," (Johnston, 2017, p. 278). Johnston also recognizes multiple "temporal channels in the human visual system," a view which is now commonplace (Johnston, 2017, p. 276). Prosser is also influenced by work in cognitive science like that of Rensink (2002), among others, on the detection of "visual transients," which draw our attention but remain at the level of subpersonal processing, and whose absence is hypothesized to play a role in change blindness (that is, our remarkable obliviousness to slowly occurring perceptual changes).

move, it does not appear to move relative to the background of the scene. That is, there is a clear sense in which it also *appears to stay still*. There is a distinct appearance of lack of motion as well as motion [emphasis original]," (Crane, 1988, p. 142). Thus, we are faced with an apparent contradiction: the cliff face next to the waterfall appears to be simultaneously in two incompatible states, i.e., moving and not moving. An illusion where we perceive objects as both moving and not moving at the same time calls out for some explanation.

The key to dissolving this paradox, it seems, is to disentangle two senses of "motion." We do not have the impression that the cliff face is *really* moving in a full-blooded sense. We do not think that the cliff face is moving in the same way the waterfall is. Neither are we confused or unsure about this, as we might be in the case of judgments of length in the Müller-Lyer Illusion.[15] It is not merely that our beliefs about what is moving remain constant, but rather that we do not directly perceive the object *as moving* at all. It is a strange experience, both like and unlike movement, but it is not the direct perception of movement *per se*. Instead, it is more plausible to say we perceive a distortion in our perception rather than the movement of an object we perceive.[16]

The "motion" detected in the illusion is easily distinguishable from its counterpart in the normal case when an object is perceived as actually changing spatial location. Le Poidevin readily admits this in his discussion of MAEs: "Ordinary perception does not seem at all like those cases of perceptual illusion where we are aware of some inconsistency," (Le Poidevin, 2007, p. 88). At best, we think of the formerly stationary object as exhibiting *motion-like* properties or having the "feel" of motion without actually being perceived as truly moving. Le Poidevin calls this the "impression of motion without any associated sense of change of relative position" (Le Poidevin, 2007, p. 89).

Thus, it is probably more accurate to interpret motion aftereffects as causing a motion-like distortion of a part of the visual field. "Pure motion," insofar as it calls this kind of phenomenon "motion," should really be considered a misnomer. In other words, instead of perceiving a change in the spatial location or extension of the objects we see, it is rather that we note a disturbance in our own vision. As we move our head, the objects *and* the surrounding context suffer the same distortion, which is not perceived as a property of the objects in themselves but instead a progressive warping of a section of the visual field, regardless of what falls within that section. The sense that the distortion is one of the visual field becomes even more salient when the gaze is shifted to a collection of objects at various distances and at various orientations, as the warping of that section of our visual field then seems quite unnatural for the objects themselves to be exhibiting individually, as it ignores

[15] For an example of the Müller-Lyer Illusion, see https://www.illusionsindex.org/ir/mueller-lyer

[16] Those that subscribe to the transparency thesis, which holds that we cannot perceive properties of experiences but only properties of the objects of them (Tye, 2014, p. 40), will perhaps not like this description. There are many reasons to doubt such a doctrine, however, for example visual noise, among other similar phenomena (Gert, 2019; see also Kind, 2003).

relative location and distance from the viewer, instead distorting the whole section in the same way.[17]

We have isolated two different perceptual scenarios with differing phenomenal characters—one illusory and one veridical. Under normal circumstances motion refers to a process whereby an object changes position in space. Evidently, this occurs in cases of perceived veridical motion like looking at an actual waterfall, but, arguably, not when we perceive the illusory case. John Frisby's (1980) description of the illusory scenario is telling in this regard: "We are still aware of features remaining in their 'proper' locations even though they are seen as moving," (ibid., 101). In other words, the relations between the parts of the objects seem to remain stable relative to each other, despite the impression of movement. What we perceive, rather than objects moving from one position to another, is the apparent warping of perceptual space. However, in occupying a part of space in the visual field that is affected by the illusion, we can say that the objects, which do not appear to change location in space, nonetheless appear to exhibit motion-like properties, in the sense of being affected by this underlying distortion.

The apparent paradox of the waterfall illusion results from conflating two perceptual phenomena under the heading "motion": (1) cases in which objects are perceived as having motion-like properties, and (2) cases where we perceive that an object has changed spatial location over time. Only cases of the second type exhibit the phenomenology of veridical motion,[18] while cases of the first might arise in both illusory and veridical cases. Although both (1) and (2) would presumably be concurrent in the case of a long exposure to the ongoing motion of an object, this is not so in the illusory case. Yet some change in position must be perceived to have experiences of motion, contrary to Arstila's purity thesis. Following Russell (1937), Le Poidevin (2017) calls the requisite kind of motion *displacement*, which cannot occur at any one particular instant because it necessarily involves a succession of differing positions. This displacement stands in contrast, we might think, with the "dynamic sensation" sought by futurist painters (Le Poidevin, 2017) and invoked by the dynamic snapshot theorists. On the other hand, accommodating perceived spatial displacement over time provides motivation for specious present views, where a conscious window can contain the successive positions of whose relations we become aware.

[17] This can be tested by viewing the illusion on a computer screen and then, rather than shifting the gaze to a stationary part of the screen, shifting to the various objects on the desk and wall near the computer, which would be at different distances and orientations from the initial stimulus.

[18] This is not to say that the converse is true, i.e., that only the phenomenology of veridical motion involves cases where we perceive that an object has changed spatial location over time. Rather, it is only to say that perceiving an object has changed spatial location over time *is* involved in veridical perception of motion. Arguably, the perception of change in position over time is also involved in non-veridical yet commonplace perceptual illusions like beta movement, a form of apparent motion that allows us to see things on screens and elsewhere as moving even though in fact there is only a sequence of frames. Apparent motion can give rise to experiences quite unlike MAEs insofar as they are much less easily distinguishable from veridical motion, if at all.

The specious present therefore allows for a more sophisticated view than the dynamic snapshot theorists can offer, involving, if we like, *both* perceived motion-like properties and the possibility of perceiving real changes in spatial location over time. Against the dynamic snapshot theorists, encoding information about motion-like properties at an instant is insufficient for any experience to be perceived as motion. As a result, on the basis of the phenomenology of MAEs, there is no reason to suppose that temporal experiences can be explained only in virtue of information encoded at an instant, as this, too, following the analogy, would be insufficient. Vector encoding, were the analogy to hold, would not result in our normal temporal phenomenology but only an unconvincing impression of it (although even this circumstance is difficult to make sense of). Such a consequence does not speak in favour of the dynamic snapshot theory.

It seems we still need *both* information encoded at a particular time *and* information that requires perception over an interval. With just the former we end up with the unconvincing illusory case, which does not seem to us like a true case of perceived motion, while with just the latter, moving objects may be perceived in staccato fashion, or at least as failing to convey a distinctive sense of movement at any given time. Indeed, the latter scenario is one of the first phenomenological inaccuracies one might associate with a snapshot theory, casting *prima facie* doubt on Arstila's punctuality thesis (2). In fact, however, such staccato experiences do occur when the motion pathway of the visual cortex malfunctions, as in the rare phenomenon of akinetopsia, or motion blindness. The next section will consider this condition in order to illustrate why visual motion and temporal phenomenology should not be conflated.

5.4 Visual Motion and Temporal Phenomenology Come Apart

Akinetopsia, or motion blindness, is instructive concerning both the mechanisms underlying motion perception and the disconnect between motion perception and temporal phenomenology. Akinetopsia reveals that normal motion phenomenology cannot be isolated to vector encoding of motion properties, although the latter, associated with activity in areas MT/V5 and MT+ of the visual cortex,[19] does play an important role in visual phenomenology (Newsome & Paré, 1988; Ajina et al., 2015). Akinetopsia also reveals that, for all the importance attached to it by the dynamic snapshot theorists, visual motion is not especially significant to our overall temporal phenomenology. For this reason, it is misguided to use the case of motion

[19] MT and V5 are different names for the same area of the extrastriate visual cortex (V1, or primary visual cortex, referring to the striate cortex). MT stands for "middle temporal," an anatomical description of the area, while V5 refers to visual processing area 5, which is a cognitive functional description. MT+ refers to area MT *plus* nearby areas, such as MST (the medial superior temporal area).

perception to make inferences about the mechanisms underlying temporal phenomenology.

Akinetopsia is a rare condition in which a subject fails to perceive motion normally, instead perceiving a stroboscopic-like effect when confronted with visual movement. In some extremely rare cases, moving objects can seem to become "stuck" or remain frozen for some period of subjective time. The key insight about these cases regarding temporal phenomenology, however, is that *subjective time itself does not freeze* in the same way as the particular moving objects in the visual field do. Rather, the subject is able to say that the object has remained stuck for some length of time. The absence of visual motion no more affects the subject's overall sense of time than the absence of colour vision.

The most studied subject to exhibit this rare condition is known as patient L.M. It is helpful to consider her case, a condensed version of which is provided by Heywood and Kentridge (2009):

> [L.M.'s] chief complaint was that she no longer saw movement; moving objects appeared 'restless' or 'jumping around'. Although she could see objects at different locations and distances, she was unable to find out what happened to them between these locations. She was severely handicapped in her daily activities, e.g. she had substantial difficulty in pouring drinks into a cup or glass, because the fluid appeared 'frozen like a glacier'; she could not see the fluid rising and was unable to judge when to stop pouring. (Heywood & Kentridge 2009, p. 24)

In the example above of liquid becoming frozen like a glacier, it is of crucial importance that only the liquid is apparently frozen. The static objects of visual experience (like the glass or kettle—in fact, everything that is not exhibiting observable motion) still persist through time, along with objects perceived through other sensory modalities, whether in motion or not. The subject's cognition also continues normally through time. She does not cease to experience time; rather, certain perceptual objects do not exhibit motion as they normally would.

The fact that L.M. has trouble judging *when* to stop pouring indicates she is aware that the liquid, which should be moving, is frozen *for* some duration. In her case, though, she must make a conscious calculation about how long a certain quantity of liquid would normally take to fill a cup in order to determine when to stop, instead of perceiving this unfold and acting accordingly, as unaffected subjects might. That the moving object is experienced as static for some interval reveals that extinguishing the phenomenology of visual motion does not prevent us from experiencing intervals of time as normal. Vector encoding, as a potential cognitive mechanism associated with visual motion, therefore appears to be a red herring when we consider that knocking out this mechanism does not grossly interfere with temporal phenomenology properly understood.

As we might expect, patient L.M. was also unable to perceive motion aftereffects like the waterfall illusion. Zihl et al. (1983) presented L.M. with a spiral-type MAE and found that she did not report experiencing the illusion at all. Nonetheless, L.M. continued to experience time, and, in fact, was able to use the timing of associated auditory cues to make inferences about visual motion, e.g., the speed and direction of vehicles nearby, to help her navigate a confusing world of staccato, yet persisting, visuals (ibid., p. 315). L.M. was thus not *temporally* impaired, but *visually* impaired. Whatever mechanism is responsible for the type of visual motion experienced in MAEs, that mechanism therefore does not seem related to temporal phenomenology in any particularly interesting sense.

Granted, the dynamic snapshot theorist is left with the logical possibility of vector encoding as the mechanism responsible for temporal phenomenology. In essence, the dynamic snapshot theorist can still say something like, "the brain appears to use vector encoding in one application, so it is possible it uses the same mechanism when it comes to other aspects of phenomenology." However, this gives us no reason to think the brain is *actually* doing this in the domain of temporal phenomenology. In fact, temporal phenomenology, which heretofore has not been connected with any particular sensory modality or area of the brain, would seem especially resistant to such an explanation. This is because, as we see in the case of akinetopsia, it is not as easily extinguished, for example by brain injury, transcranial magnetic stimulation, or psychopharmacological intervention, as other, less fundamental aspects of phenomenology such as visual motion. Indeed, the extinction of temporal phenomenology seems impossible without eliminating consciousness entirely.

Besides the disconnect between visual motion phenomenology and temporal phenomenology more generally, akinetopsia also makes clear that normal motion phenomenology is not generally a matter of one localized mechanism, like vector encoding. This becomes clear when we consider that, even though the akinetopsic patient L.M. suffered from the near total elimination of normal functioning in area MT/V5, motion phenomenology was *not* completely eliminated. Heywood and Kentridge elaborate: "Although L.M. has been dubbed 'motion-blind', she retains rudimentary movement vision, e.g. she can discriminate speed and direction of motion of high-contrast gratings at low speeds," (Heywood & Kentridge, 2009, p. 25).

The closely related underlying neurophysiology of akinetopsia and MAEs indicates the former condition is much like a chronic inversion of motion aftereffects. Akinetopsia typically results from lesions to visual cortex area MT/V5 and can also be induced by transcranial magnetic stimulation (TMS) to this area (Beckers & Zeki, 1995). MT/V5 is active in cases of illusory motion like the waterfall illusion,[20] as well as in cases where subjects perceive "implied motion," for instance in comic strips (Heywood & Kentridge, 2009, p. 25). In addition to inducing akinetopsia, when TMS is applied to MT/V5 it is capable of eliminating motion aftereffects in normal subjects (ibid.). Naturally, MT/V5 is the proposed site of Prosser and Arstila's vector encoding mechanism. In the case of the akinetopsic subject this area is essentially deactivated.

One conclusion we can draw from the neurophysiological analysis of akinetopsia is that visual motion is not the result of one simple process. Information perceived over a span of time remains relevant to our normal phenomenology of motion and is not parasitic or secondary to the operation of V5/MT, meaning the purity

[20] The motion-like properties of static objects during MAEs is associated with false signals arising from specialised motion detection neurons that "suffer a reduction in responsiveness" as a result of sustained exposure to the moving stimuli (Anstis et al., 1998, p. 111).

thesis looks increasingly dubious. This becomes even clearer when we consider that important perceptual features are retained in akinetopsic patients, including some degree of motion perception:

> [...] Akinetopsia provides a clear example of *selective* loss of phenomenal consciousness for visual motion [emphasis added]. Patient L.M. [...] retained the ability to use biological motion cues which are probably processed by brain areas distinct from those concerned with processing motion of rigid bodies and global motion of a scene. (Heywood & Kentridge, 2009, p. 25)

Progress in understanding the areas of the brain responsible for vision reveals that besides processing properties like shape and colour, the brain has a specialised "visual motion pathway."[21] According to Newsome and Paré (1988, p. 2201), this pathway "originates in striate cortex and terminates in higher cortical areas of the parietal lobe," processing and giving rise to experiences of the motion-like properties of objects (see Rokszin et al., 2010 for a detailed neurophysiological description). As mentioned, crucial parts of this pathway, most notably processing in area MT/V5, can be severely disrupted for patients with akinetopsia and it is also this pathway that is affected in visual motion illusions, of which the waterfall illusion is one example. However, this pathway is more complex than the dynamic snapshot theorist appreciates, involving several stages, multiple cell types, multiple streams, and more areas than V5/MT alone (Rokszin et al., 2010). More importantly, this pathway is not responsible for any aspect of temporal phenomenology, as becomes apparent when we consider that its malfunction does not disrupt akinetopsic patients' overall sense of time continuing to flow. What we see instead is that disruption here leads to improbable, abnormal visual perceptions of the way objects move through space—not time. For these reasons we should be sceptical of claims that the alleged vector encoding mechanism of V5/MT responsible for some aspects of visual motion perception is somehow generalizable to time-consciousness.

One might think of the motion detection pathway as providing a quick and easily accessible way of encoding potentially life-threatening information, like the speed and direction of an incoming predator, which is quickly registered by these neurons without having to consider a large change in location over a span of time.[22] A tiger, to follow this example, can be simply attributed motion and direction without having to perceptually register a substantial change in spatial location, as might be required for the full-blooded perception of motion (stripes and all) that we would normally have upon seeing a tiger run for some length of time. Arguably, the length of time necessary for the normal, fleshed out perception may not always be enough, in which case the visual motion pathway could prove highly adaptive. The neurophysiology of MAEs therefore indicates the waterfall illusion phenomenon and its

[21] See Johnston (2017) for an accessible discussion of the neurophysiology.

[22] A TMS study by Beckers and Zeki (1995) has indicated that "perceptually effective visual motion signals reach V5 at or before 30 ms and reach VI [V1] at or before 60 ms," and the authors concluded on this basis that the brain employs both a fast (direct to V5) and a slow (through V1) pathway when processing motion (ibid., p. 49).

associated physical substrate are the result of unique adaptations likely unrelated to aspects of temporal phenomenology like succession and continuity, which we should continue to think are related to more primitive, global features of consciousness, contrary to the fourth thesis of the dynamic snapshot theory.

As our look at akinetopsia shows, it is certainly not the case that encoding of motion-like properties in the visual cortex just *amounts to* our experience of time. We should also be wary of the claim that similar processes would be at work in all aspects of temporal phenomenology. One reason to resist such an extrapolation is that, unlike in the motion case, there are no identifiable neural populations or brain areas picking up on fundamental features of temporality like succession from sensory stimuli in the same way that motion can be picked up from vision. We also do not find cases of patients presenting with a loss of fundamental aspects of temporal phenomenology like experiential succession (i.e., one thing happening after another). If it were possible for brain lesions to result in disturbances of this kind, we might think the processes underlying temporal phenomenology were localized in specific areas. These cases might convince us that temporal structure is not a global, fundamental feature of consciousness, but a bolt-on module or a series of modules. However, such cases are simply not forthcoming.

So long as a subject is conscious, there seems to be a basic experience of time passing. Barry Dainton (2010) aptly illustrates the ubiquity of the sense of passage through the example of lying in a deckchair looking at the sky, as well as examples from other sensory modalities:

> For some moments now you have been staring at an empty region of blue sky and nothing has changed. Your inner monologue has (if only briefly) ground to a halt, you have seen no movement, your visual field is filled with an unvarying expanse of blue. But is your consciousness entirely still or frozen? Have you come to a complete stop? No. Throughout this period you remain conscious, and conscious of the blue presence *continuing on*; you have a (dim, background, passive) awareness of the blue constantly being renewed from moment to moment. This passive awareness of continuation and renewal is perhaps more vivid in the case of auditory experience. Imagine hearing a sustained but unwavering note played on a cello: you hear a continuous and continuing flow of sound. This feature—call it "immanent phenomenal flow"—is possessed by all forms of experience (think of the burning sensation on the tongue caused by biting on a chilli pepper), and is a dynamic feature of experience that is independent of changes of the ordinary qualitative sort (the chilli-induced burning is felt as *continuing on* even when its intensity and qualitative character remains constant).

This ubiquitous sense of flow is *not* dependent on movement or change in vision in the same way as it is not dependent on changes to the qualitative feel of burning chilis. This sense of flow does not lend itself to description in the vector-like terms of trajectory or velocity and does not easily analogize to the movement of objects through space. Confusing temporal properties for spatial properties, which may be detected by the visual system, seems to be at the heart of the conceptual error that has led to the mistaken association of vector encoding with temporal phenomenology.

The ubiquity of the sense of passage would seem to suggest that it cannot result from disparate encapsulated mechanisms and that it is more likely a feature of the process or processes responsible for consciousness in general. For Arstila, a "second-order motion processing mechanism" is responsible for the allegedly "temporal" phenomenology in the waterfall illusion (Arstila, 2018, p. 291). However, according to Arstila, similar mechanisms *also* exist for every other aspect of temporal phenomenology. For example, similar, though separate mechanisms must be found for the experiential succession in the context of thinking and deliberating, the entertaining of beliefs and desires, interoception, proprioception, emotion and mood states, etc. Not only is there no evidence of analogous mechanisms here, but the dynamic snapshot appeal to vectors (representing direction and velocity) just does not coherently apply in many of these cases. We may possibly diagnose the confusion here as resulting from something like an unwarranted spatialization of time. We might also consider that, while quickly registering the motion of a tiger would be of the utmost importance to survival, there is no pressure to develop a similar mechanism for encoding information about the succession of thoughts at an instant, were this even possible. Besides sidestepping these issues, appealing to a general mechanism has the benefit of accounting for why people never seem to go "timeless" the way people go "blind". On this view temporality is indispensable to consciousness because it is inextricably bound up with the processes responsible for it.

5.5 Departing from Local Mechanisms

A certain strand of philosophy dating back at least to Kant (1787) has contended that temporality is really a necessary precondition for conscious experience. For Kant, "Time is [...] given *a priori*. In it alone is actuality of appearances [phenomena] possible at all. Appearances may, one and all, vanish; but time (as the universal condition of their possibility) cannot itself be removed," (Kant, 1787, p. A31). The necessity of time for experience can be supported by appeal to the inconceivability of experience without temporal structure. This inconceivability is not just a failure of imagination on our part. Experiences as we know them are essentially in time and occur over time—an experience that was not so would not be worthy of the name, besides being inconceivable.

Given that consciousness is essentially temporal, it is plausible that temporality is an inherent feature of the widely distributed processes responsible for subjective experience generally. A recent proposal from Hohwy et al. (2016) runs with this notion, placing temporality at the heart of the predictive mechanism that Bayesian-brain-style information processing accounts see as responsible for consciousness (and much else besides). For Hohwy, Paton, and Palmer, it is the continuous update and replacement of our cognitive system's predictions about the hidden causes of our sensory stimuli (our models of the world) in light of new information from an ever-volatile environment that results in a feeling of rolling along through time (ibid.). Such a theory presents temporality as a deep feature, found throughout a multilevel mechanism of cognition, which is inherited by any conscious event.

Isolating subjective temporality to one particular neural or cognitive mechanism, e.g., vector encoding, associated with a particular modality or phenomenon is too restrictive. Rather, it is in global processes that we may have more luck identifying the neural and cognitive correlates of temporal phenomenology in a way that illuminates why it is so fundamentally inescapable for a conscious being.

If this is right, then the attempt to connect temporal phenomenology to relatively local processes like the mechanisms at work in visual motion processing is a deeper flaw than it first appears. The mistake becomes clearer when we look at the implications of the dynamic snapshot view, were we to accept it. Suppose we grant that motion can be perceived in a robust way solely via the vector encoding mechanism that illusory motion from MAEs is meant to demonstrate. Instead of explaining temporal phenomenology, such a position really only postpones explanation. This is because the phenomenology of visual motion, if determined by vector encoding at an instant, must *still* appeal to an overarching temporal phenomenology for the illusion, or indeed anything, to be experienced at all.

To elaborate, let us assume, as per the dynamic snapshot theory, that perceived motion of the illusory kind is sufficient for the phenomenology of visual motion and there is no direct perception of change in spatial location over an interval of time. Despite this, the experience of the illusory motion persists through time and must do so in order for us to have it. When faced with a static object perceived as having motion-like properties, it is not the case that our experience has stopped, is static, or is confined to an instant. The organism's overall consciousness does not become instantaneous (in other words: confined to an instant or solely about instantaneous states of affairs) just because we are experiencing an unmoving object. For the experienced illusory object (the cliff face, say) to appear static in space, but with motion-like properties, it must exhibit persistence over some length of time. This temporality is inescapable; events continue to succeed one another over time, even if the content remains the same.

Whether in the illusory or non-illusory case, while the cliff next to the waterfall is either stationary or apparently exhibiting motion-like properties, our experience of that object must continue through a window of time for us to experience that object as either static *or* moving. It doesn't give the dynamic snapshot theorists any further ammunition that the cliff-face can be seen to exhibit motion-like properties, as the perception of any properties whatsoever still requires an extended experience in the way that the competing, "specious present" models of temporal phenomenology identify. These views agree, contrary to the purity thesis, that the experiential present must be an interval of time for us to enjoy the phenomenology we are familiar with, like succession and continuity.

Against the idea of a present window, Arstila cites the frequent disagreement over lengths of the temporal interval to suggest that no such specious present exists. However, insofar as an interval of any length is indicated at all by empirical studies, it cannot be concluded that a snapshot view is superior to a specious present view. Rather, it would be more reasonable to maintain there may be some inherent task-dependent flexibility in the duration of the specious present, as James (1890) long ago suggested.

5.6 Conclusion

I have argued that the waterfall illusion and other motion aftereffects do not support the dynamic snapshot theory. The misappropriation of motion aftereffects in support of the dynamic snapshot theory results from a faulty interpretation of these illusions. Furthermore, the connection between visual motion and temporal phenomenology turns out to be spurious, meaning that, despite the role of vector encoding mechanisms in the former, it is an unjustified leap from that mechanism to the assertion of a similar mechanism underlying temporal phenomenology. Consequently, motion aftereffects fail to undermine the specious present as snapshot theorists might hope, while countervailing considerations remain as strong as ever.

We do not have good reason to think that things like succession, continuity, duration, and other aspects of time-consciousness result from or are related to similar mechanisms as the vector encoding that may occur in area V5/MT when we are confronted with the waterfall illusion. This should be a cautionary tale for theories seeking to provide explanations of temporal phenomenology. Evidence for or against models of temporal phenomenology will not be found at the level of the neural mechanisms responsible for any one specific sensory modality. Rather, models of temporal phenomenology should be intimately related to models of subjective consciousness generally, as temporal phenomenology is a precondition for having subjective experiences in the first place and not an "additional" feature of consciousness. Chapter 7 will take up this line of thought in much greater detail, developing the view that it is in the basic operation of the predictive engine of the brain that temporal phenomenology begins, but already the motivation for locating the explanation in such low-level and non-localized processing should be clear. In the meantime, in the next chapter, I will provide a positive proposal of a specious present view that can better account for temporal phenomenology and which can also be readily grounded in a plausible information processing architecture.

References

Addams, R. (1834). An account of a peculiar optical phenomenon seen after having looked at a moving body. *London and Edinburgh Philosophical Magazine and Journal of Science, 5*, 373–374.

Ajina, S., Kennard, C., Rees, G., & Bridge, H. (2015). Motion area V5/MT+ response to global motion in the absence of V1 resembles early visual cortex. *Brain, 138*(1), 164–178.

Anstis, S. M., Verstraten, F. A. J., & Mather, G. (1998). The motion aftereffect: A modern perspective. *Trends in Cognitive Sciences, 2*(3), 111–117.

Arstila, V. (2018). Temporal experiences without the specious present. *Australasian Journal of Philosophy, 96*(2), 287–302.

Bach, M. (2004). *Flash-Lag Effect*. Retrieved December 1, 2019, from https://michaelbach.de/ot/mot-flashLag/index.html

Beckers, G., & Zeki, S. (1995). The consequences of inactivating areas V1 and V5 on visual motion perception. *Brain, 118*(1), 49–60.

Blakemore, C. (1973). The baffled brain. In R. L. Gregory & E. H. Gombrich (Eds.), *Illusion in nature and art* (pp. 8–47). Duckworth.
Broad, C. D. (1923). *Scientific thought: A philosophical analysis of some of its fundamental concepts*. Kegan Paul.
Chuard, P. (2011). Temporal experiences and their parts. *Philosophers' Imprint, 11*(11).
Crane, T. (1988). The waterfall illusion. *Analysis, 48*(3), 142.
Crick, F., & Koch, C. (2003). A framework for consciousness. *Nature Neuroscience, 6*, 119–126.
Dainton, B. (2010). *Time and space* (2nd ed.). Acumen.
Dainton, B. (2017). Temporal consciousness. In E. N. Zalta (Ed.), *Stanford Encyclopedia of Philosophy* (Fall 2017 Edition), https://plato.stanford.edu/archives/fall2017/entries/consciousness-temporal/
Dennett, D. (1993). *Consciousness explained*. Penguin.
Frisby, J. P. (1980). *Seeing: Illusion, brain, and mind*. Oxford University Press.
Gert, J. (2019). Transparency, representationalism, and visual noise. *Synthese, 198*(7), 6615–6629. https://doi.org/10.1007/s11229-019-02480-7
Gregory, R. L. (1966). *Eye and brain: The psychology of seeing*. Weidenfeld and Nicolson.
Heywood, C., & Kentridge, R. W. (2009). Akinetopsia. In T. Bayne, A. Cleeremans, and P. Wilken (Eds.), *The Oxford companion to consciousness*. Oxford University Press.
Hoerl, C. (2017). On the view that we cannot perceive movement and change: Lessons from Locke and Reid. *Journal of Consciousness Studies, 24*(3–4), 88–102.
Hohwy, J., Paton, B., & Palmer, C. (2016). Distrusting the present. *Phenomenology and the Cognitive Sciences, 15*, 315–335. https://doi.org/10.1007/s11097-015-9439-6
James, W. (1890/1980). *The principles of psychology*. Henry Holt.
Johnston, A. (2017). Perceiving visual time. In I. Phillips (Ed.), *The Routledge handbook of philosophy of temporal experience*. Routledge.
Johnston, A., McOwan, P. W., & Benton, C. P. (1999). Robust velocity computation from a biologically motivated model of motion perception. *Proceedings of the Royal Society of London B: Biological Sciences, 266*(1418), 509–518.
Johnston, A., McOwan, P. W., & Buxton, H. (1992). A computational model of the analysis of some first-order and second-order motion patterns by simple and complex cells. *Proceedings of the Royal Society of London B: Biological Sciences, 250*(1329), 297–306.
Kant, I. (1787/2007). *Critique of Pure Reason*, revised 2nd ed. Translated by N. Kemp Smith. Palgrave Macmillan.
Khoei, M. A., Masson, G. S., & Perrinet, L. U. (2017). The flash-lag effect as a motion-based predictive shift. *PLoS Computational Biology, 13*(1), e1005068. https://doi.org/10.1371/journal.pcbi.1005068
Kind, A. (2003). What's so transparent about transparency? *Philosophical Studies, 115*(3), 225–244.
Le Poidevin, R. (2007). *The images of time: An essay on temporal representation*. Oxford University Press.
Le Poidevin, R. (2017). Motion and the futurists. In I. Phillips (Ed.), *The Routledge handbook of philosophy of temporal experience* (pp. 315–325). Routledge.
Lee, G. (2014). Extensionalism, atomism, and continuity. In N. Oaklander (Ed.), *Debates in the metaphysics of time*. Bloomsbury.
Locke, J. (1690/1975). In P.H. Nidditch (Ed.), *An essay concerning human understanding*. Clarendon Press.
Macpherson, F., & Baysan, U. (2017). Waterfall illusion. In F. Macpherson (Ed.), *The Illusions Index*. Retrieved December 1, 2019, from https://www.illusionsindex.org/ir/waterfall-illusion
Mather, G., Verstraten, F., & Anstis, S. (1998). *The motion aftereffect: A modern perspective*. MIT Press.
McKenna, C. A. (2021). Don't go chasing waterfalls: Motion aftereffects and the dynamic snapshot theory of temporal experience. *Review of Philosophy and Psychology, 12*(4), 825–845.

Miller, K. (2019). Does it really seem to us as though time passes? In V. Arstila, A. Bardon, S. E. Power, & A. Vatakis (Eds.), *The illusions of time: Philosophical and psychological essays on timing and time perception*. Palgrave Macmillan.

Newsome, W. T., & Paré, E. B. (1988). A selective impairment of motion perception following lesions of the middle temporal visual area (MT). *The Journal of Neuroscience, 8*(6), 2201–2211.

Piper, M. S. (2019). Neurodynamics of time consciousness: An extensionalist explanation of apparent motion and the specious present via Reentrant oscillatory multiplexing. *Consciousness and Cognition, 73*, 102751.

Prosser, S. (2016). *Experiencing time*. Oxford University Press.

Prosser, S. (2017). Rethinking the specious present. In I. Phillips (Ed.), *The Routledge handbook of philosophy of temporal experience*. Routledge.

Reid, T. (1785/2002). In D. Brookes and K. Haakonssen (Eds.), *Essays on the intellectual powers of man*. Edinburgh University Press.

Rensink, R. A. (2002). Change detection. *Annual Review of Psychology, 53*(1), 245–277.

Rokszin, A., Márkus, Z., Braunitzer, G., Berényi, A., Benedek, G., & Nagy, A. (2010). Visual pathways serving motion detection in the mammalian brain. *Sensors, 10*(4), 3218–3242.

Russell, B. (1937). *The principles of mathematics* (2nd ed.). Cambridge University Press.

Shardlow, J. (2019). Minima sensibilia: Against the dynamic snapshot model of temporal experience. *European Journal of Philosophy, 27*, 741–757.

Tye, M. (2014). Transparency, qualia realism and representationalism. *Philosophical Studies, 170*(1), 39–57.

Wittmann, M. (2017). *Felt time: The science of how we experience time* (E. Butler, Trans.). MIT Press.

Wohlgemuth, A. (1911). *On the after-effect of seen movement* (Vol. 1). Cambridge University Press.

Zihl, J., von Cramon, D., & Mai, N. (1983). Selective disturbance of movement vision after bilateral brain damage. *Brain, 106*(2), 313–340.

Chapter 6
ERA: The Extensional-Retentional Analysis

This chapter will argue not only that extensionalism and retentionalism are compatible positions, but that the most sensible understanding of the phenomenology of the present will appeal to features of both of them. I do not mean to say here that extensionalism and retentionalism should be superimposed upon each other as redundant explanations of the same phenomena, but rather that *different* temporal phenomena should be accounted for in different ways. Contrary to most historical and contemporary positions on the matter, I think it is far from obvious that a singular analysis is appropriate for such distinctive phenomena as succession, duration, and the temporal ordering of events. The complexity and diversity of the phenomena under investigation would seem instead to call for a pluralistic explanatory strategy. My positive proposal is that a specific way of drawing on the explanatory resources of extensionalism and retentionalism, which I will call "ERA" for brevity,[1] better accounts for temporal phenomenology than either retentionalism or extensionalism alone, and moreover arises naturally out of the cognitive apparatus responsible for consciousness. This chapter will describe ERA and the reasons for adopting it, while the next chapter will show how our temporal phenomenology is ultimately rooted in the way that our predictive minds operate.[2]

---

[1] ERA stands for "extensional-retentional analysis". However, the proposal is in reality several *different* analyses applied to different explanatory targets, rather than a single hybrid analysis of temporal phenomenology. I am leaning on the traditional taxonomy only insofar as the classic models can help to illustrate the way ERA is supposed to work. If the labeling does not seem exactly appropriate, then I invite readers to dispense with the taxonomy altogether and consider only what the proposal involves.

[2] ERA and TIER, discussed in the next chapter, are really two sides of the same coin. In fact, it was the connection between cognitive processing and phenomenology that initially inspired this book. It is therefore best if these two Chaps. 6 and 7 are considered together so the context is as clear as possible. Some may even find it more illuminating to read the next chapter before this one because this was how these ideas occurred to me chronologically, but narratively I felt that moving from phenomenology to implementation would make for a smoother transition.

C. A. McKenna, *Constructing the Present*, Studies in Brain and Mind 24,
https://doi.org/10.1007/978-3-031-89571-5_6

I will begin by showing how classical forms[3] of retentionalism encounter a potentially infinite regress that risks undermining the retentional accounting of the specious present, while a pluralistic approach like ERA does not face similar concerns. Relatedly, I suggest that ERA evades the unpalatable implication of the classical retentionalist picture that there is no principled reason why the specious present might not extend indefinitely, without a non-arbitrarily defined boundary for the presently experienced content. We can overcome these issues by acknowledging that the actual temporal extension of intentions really is, as extensionalists claim, explanatorily relevant for the phenomenal character of our experience of the present. I then make the case that a pluralistic explanatory strategy evades what Barry Dainton (2017)[4] has identified as the "extensional" and "retentional simultaneity problems", which stem from oversimplifications made by each model individually in accounting for specious present. I will then argue that an extensional-retentional analysis strikes the right balance between parsimony and phenomenal accuracy while still accommodating temporal illusions through appeal to intentional content. Ultimately, I conclude that such an analysis provides a more satisfactory understanding of experiential succession, temporal variability, and the duration of the specious present than the standard accounts can offer on their own.

While the simplest way of conjoining extensionalism and retentionalism may be to lean on both to account for any given stream of consciousness, e.g., by maintaining an extensionalist overlap model *and* a Husserlian retentional complex, the upshot of doing so would be an unrealistic surfeit of temporal experience.[5] Such a "superimposition" of views is not what I have in mind here. Instead, ERA represents a particular *pluralistic explanatory strategy* and is not a simple hybrid model.

In my view, certain phenomenal features of our experience of the present are amenable to extensionalist analyses, while others are more amenable to retentional analyses. This pluralism should be seen as a positive feature, not a weakness, of an account of complex mental phenomena that share little in common. In particular, according to ERA, experiential succession and the duration of the specious present are better accounted for by extensionalism. More precisely, ERA appeals to two flavors of extensionalism: "relational" extensionalism to account for succession and "inheritance" extensionalism to account for duration. Note that neither of these commitments involve a commitment to any singular understanding of how perception works in general, but are restricted to the target phenomena, which I do not think should be lumped into the same basket as, say, color or object perception. In part, this is because, as we have seen, such fundamental temporal phenomena are enabling conditions for other kinds of perception and therefore should not be assumed to be the same kind of thing. Thus, in endorsing relationalism to account

[3] I.e., those that are indebted to Husserl's original formulation of retentionalism (Husserl, 1917).

[4] The names of these problems were coined by Barry Dainton in an older version of his Stanford Encyclopedia of Philosophy article on "Temporal Consciousness" (Dainton, 2017) but are not found in the most recent edition as of writing (Dainton, 2023).

[5] I would like to thank an anonymous reviewer for pushing me to make this clearer.

for succession, I intend to remain agnostic about the truth of relationalism about perception in general (and likewise for inheritance).[6]

As a pluralistic explanatory strategy, ERA's commitments are to specific propositions with precisely restricted domains. Thus, it is perfectly consistent for ERA to maintain that (1) experiential succession should be accounted for in a relational way; (2) duration should be accounted for by appeal to inherited temporal properties; (3) event ordering should be determined by properties[7] of content-bearing intentions; and finally (4) the ongoing generation of this diachronically unified succession of changing relations to experiential objects,[8] which is underwritten by the flux of expectations (understood, if one prefers, as intentions), results in the construction of the present such that time-consciousness is best understood overall from an adverbial perspective, i.e., as the *way* we perceive and not what we perceive.

ERA seeks to avoid appealing to intentionality in accounting for these most basic aspects of temporal phenomenology, not least because intentionality already requires a temporally structured subject. At the same time however, features of experience like the temporal relations of experiential events to each other and their variability under illusory conditions are seemingly more explicable in retentional terms. Importantly though, it is not the intentional content but rather the *kind of intention* that ERA takes to specify the temporal relations between events. This is in keeping with the general view that in terms of fundamental phenomenology, time is not itself an object of experience and not explicable in terms of just more content, even if, derivatively speaking, it can be represented.

Ultimately, attention to the finer-grained details of our phenomenology and cognitive processing should cast doubt on our ability to provide a satisfactory uniform accounting of the phenomena under investigation. The reasons for favoring a pluralistic strategy like ERA and the details of the view itself will also become much clearer in light of the next chapter, which will look at how the predictive approach in cognitive science dovetails with an analysis of this kind.

---

[6] As a pluralistic explanatory strategy, ERA's commitments are to specific propositions with precisely restricted domains. This is why the positions are not incompatible. Thus, it is perfectly consistent to maintain that (1) succession should be accounted for in a relational way, (2) duration should be accounted for by appeal to inherited temporal properties, and (3) event ordering should be determined by properties of content-bearing intentions, and finally (4) the ongoing generation of this diachronically unified succession of changing relations, which is underwritten by the flux of expectations (understood as intentions), results in the construction of the present such that time-consciousness is best understood overall from an adverbial perspective, i.e. as the way we perceive and not what we perceive.

[7] Properties like what role the intention/expectation plays in the cognitive system and how it is related to others.

[8] The nature and ontological status of such objects can remain unspecified here. These things are not as relevant to the temporal phenomenology as the relational process itself, in my view.

6.1 Two Problems for Classical Retentionalism

According to the classical Husserlian form of retentionalism, experience is composed of retentions, retentions of retentions, as well as retentions of these, and so on, in a process of modification that continues "*ad infinitum*" (Husserl, 1917, p. 130).[9] However, if the process is naturally *ad infinitum* as Husserl says, then it is unclear why there should be any end, however dynamic, flexible, or fuzzy, confining retentional content to an experiential present lasting any particular amount of time. In principle, the intentions posited by retentionalism would extend forever outwards from the specious present in both directions unless there is some additional limiting factor(s). Husserl himself recognized this problem in a footnote to his diagram of the retentional model: "No notice is taken in the diagram of the limitation of the temporal field. No end to retention is provided for therein, and, ideally at least, a form of consciousness is possible in which everything is retentionally retained," (Husserl, 1917, pp. 52–53, n. 8). Nonetheless, Husserl thinks that "the same complex [of intentions] continuously undergoes a modification until it disappears, for hand in hand with the modification goes a diminution which finally ends in imperceptibility," (Husserl, 1917, p. 52). He then claims that "the temporal field always has the same extension," (ibid.) and is analogous to space in this way. Why and how such a diminution occurs and why the field would remain finitely bounded to any particular extent, remains unclear. Husserl may have been content merely to describe it as such. However, I will argue that, when faced with the question of *why*, any retentionalist appeal to brute facts like brain processes will turn out to undermine the motivation for the view itself.

There are really two problems in this area. The first is the problem of how to make sense of potentially infinitely *iterated* content (retentions of retentions etc., *ad infinitum*), and the second is the problem of how to make sense of the potentially infinite *expanse* of temporal content constituting our present experience. We can call the first the "Infinite Echo Problem" and the second the "Problem of the Infinite Present". On its own, the retentional model itself does not provide for why either of these things should have finite limits, even though, manifestly in our experience, there are limits. Each experience of ours does not reverberate infinitely, and neither does the interval we take to be now extend indefinitely over our entire lives or more. This inconsistency between what the model suggests in principle and what and our experience tells us demands resolution.

Of course, there simply are physiological constraints on brain activity and it therefore may seem appealing for a retentionalist to lean on these to overcome both problems. Doing so would mean, for instance, that the duration of certain physiological processes determines, either directly or indirectly, the precise finite interval of the experienced present and the actual degree of intentional nesting. Crucially, however, this appeal to the properties of the supposed content *bearers* betrays the

---

[9] This description simplifies Husserl's view by leaving out other kinds of intentions, but those details are unimportant for current purposes.

typical intentionalist, and *a fortiori,* the typical retentionalist, conviction that it is the contents of intentions alone that fully determine the phenomenal character of experience. A response like this therefore abandons classical retentionalism's core commitment to the explanatory sufficiency of intentionalism.[10] Moreover, if we abandon intentionalism and accept that properties of the content-bearer *are* relevant to the phenomenal character of experience, this then allows all sorts of explanations of that kind, including classical extensionalism. The impetus for preferring retentionalism over extensionalism on the basis of prior intentionalist commitments then evaporates.

To reiterate: an explanatory appeal to brute facts or physical limitations is an appeal the strict retentionalist seemingly cannot make without either violating their intentionalist convictions about the mind writ large[11] or ignoring the physical facts and being arbitrary about brute boundaries. But if we reject intentionalism (and I argue we should), then there is no problem making appeals to brute facts or physical processes to account for other aspects of our phenomenology. It is not so much that retentionalism *necessarily entails* the problems of unconstrained infinities, but that this problem requires some explaining away which is not obviously forthcoming if we remain committed to intentionalism. And once we decide to depart from intentionalism by entertaining such explanations, other, arguably more straightforward, options open to us.

As the architect of the phenomenological method, which brackets ordinary scientific investigation, Husserl himself would not appeal to brain processes to solve these problems. However, those that think what the brain is doing is relevant to the character of our experience might well be sympathetic to such an appeal. Indeed, merging retentionalism with empirical cognitive science and neurophysiology has become a minor cottage industry under the banner of what Francisco Varela (1996) has called "neurophenomenology."[12] Such projects appear to be compatible with the more ecumenical view I am defending here and, in some cases, may even be committed to something like it already. This is perhaps most obvious in the case of Varela (1999, 2000), who privileges the timescale of "transient neural assemblies" in determining the temporal window of experience, while also committing to retentionalism in principle. More recently, Wanja Wiese (2017), building on Rick Grush's

[10] The retentionalist commitment to intentionalism ultimately derives from Husserl's mentor Brentano and can be understood as the notion that intentional content determines the phenomenal character of experience (Brentano, 1874/2015, pp. 88–89).

[11] Note that where one stands on these issues often plugs into convictions about the overall nature of perception. For example, representationalism/intentionalism about perception as a whole appears most compatible with retentionalism, while extensionalism may appear more compatible with naïve realism. The larger position can motivate and constrain the position on temporal phenomenology. However, as I have stressed elsewhere, temporal phenomenology is rather unlike run-of-mill experiences. Shoehorning it into a theory originally developed with other sensory phenomena in mind fails to recognize the ways in which time is special, in my view.

[12] Examples include diverse work by Varela (1999), Van Gelder (1996), Lloyd (2004), and Grush (2005), among others. See also Thompson (2010, Chap. 11). See also Northoff (2016, 2023) for a more extensionalist-leaning foray into neurophenomenology.

computationalist "trajectory estimation model" (2005), maintains that "the content of the specious present is not best conceived as a linear stream of events, but rather as a hierarchy of temporal wholes," (2017, p. 3). Wiese thinks that his account is "compatible with all models that embrace the view that the content of momentary conscious experience comprises an interval (which is common ground between retentional and extensional models," (ibid., p. 4). Wiese's view, which adapts the predictive processing framework[13] to the temporal domain, represents a promising direction that appeals to the intentionalism of Husserl while maintaining the potential explanatory relevance of the structure of information processing and the neurophysiological underpinnings of that structure.[14] This idea will be discussed in greater detail in the next chapter.

6.1.1 Extensionalism as Inheritance

For their part, extensionalists do not face Problem of the Infinite Present because they typically reject the sharp distinction between properties of experiential content and properties of experiential content bearers. That is to say, they reject the "content/vehicle" distinction, at least within the domain of temporal phenomenology.[15] Because they do not think intentional content determines the duration of the specious present, extensionalists face no problem of infinite intentional expanse. Experience simply lasts a certain amount of time and it is this fact that determines in some way the duration of the specious present. In eschewing intentional content the extensionalist also does not face an infinite echo. For an extensionalist there is no problem appealing to brute facts or physiological processes because this is in keeping with their *modus operandi*—extensionalists reject the notion that temporal experience should be explicated as *just more content*. Nevertheless, for the extensionalist, a new question arises: why and how should properties of the experience itself or the processes responsible for it have any explanatory purchase on the phenomenal character of our experience?

A reply that appeals to common-sense intuitions might be: why would they not? Even accepting this idea uncritically though (and we won't), we still need a more thorough account of how exactly such an explanation is supposed to work. One way it might work is that temporal properties just fall out of the form of the experiential content bearers such that those properties are *inherited*[16] by the experience. We can

[13] See Hohwy (2013) for a wide-ranging overview of predictive processing and Hohwy et al. (2016) for a predictive processing inspired account of temporal experience, in this case specifically with regards to experienced "flow."

[14] For more on Wiese's and Grush's models in particular see Chap. 7, Sect. 7.4.

[15] See Hoerl (2013), esp. Section 2 (p. 383 in particular), for a fair-handed discussion of the debate between extensionalists and retentionalists and what it consists in.

[16] See Phillips (2014a, 2014b) for naïve realist extensionalist defenses of what Phillips calls the "inheritance principle," which holds that "for any temporal property apparently presented in per-

call this the "resemblance" view of extensionalism (after Hoerl, 2013, p. 381). In fact, there are many instances where the properties of a representational vehicle appear to have a *prima facie* obvious bearing on the properties of the content of the representation. While in principle representations can represent their content in any arbitrary way (they can be strings of arbitrary symbols like these words, for instance), nonetheless in some cases the connection is apparently non-arbitrary. In pictorial representations, for example, certain properties of the vehicles—their colors, their shapes, perhaps their texture—might have direct relevance to the content represented. So, while the word "green" does not need to be green to represent the color green, a pictorial representation of something green, like a pine tree, *can* represent the color properties of its content in virtue of itself being green; similarly, we might consider a hieroglyphic representing a bird in virtue of having a broadly similar shape to a bird. Though contingent, the connection in these examples is non-arbitrary—the reason the bird-shaped hieroglyph represents bird-content is, at least to some extent, the isomorphism between the representational vehicle (the symbol) and its content.

Could not something similar occur with temporal properties? Tree rings, for instance, can be considered representations of temporal content, and it so happens that the quantity and arrangement of successive rings—a property of the vehicle—determines the temporal content that can be read off the representation, i.e., the number of years the tree has existed. One might think such representations would be the simplest solution to our problem with experience. Granted, this dendrochronological example is not perfectly analogous to experience because in the tree ring case non-temporal properties (e.g., the spatial arrangement and number of the rings) represent temporal properties. However, there remains a significant relationship between the (temporal) properties of the content and the (spatial) properties of the vehicle. In the experiential case, we might think temporal properties like duration are represented by temporal properties of the *same kind*. More bluntly, the actual duration of the unfolding experience might simply *be* the duration felt. On a physicalist account, the duration of this unfolding would ultimately be determined by the duration of a corresponding physical process. We might therefore say the temporal properties of the vehicle would determine (at least some of the) character of experience, either directly or indirectly.

It is of course unlikely that there is any experience associated with tree rings and it would be absurd to claim that the mere fact that something is represented provides any explanation for why experience is felt the way it is. In the case of *experiential properties* though we assume that whatever sort of representation is involved is the kind where the content is already *experiential* content. How and why exactly that content becomes experiential is a different question which is not the target here. Given all this, we might see the above examples as providing an opening for the

ceptual experience, experience itself has that same temporal property," (Phillips, 2014a, p. 131). Phillips also notes that adherence to the principle is not "obligatory" for the extensionalist (Phillips, 2014a, p. 133).

possibility that vehicle properties can have explanatory relevance for the associated phenomenal character of experience by way of inheritance.

6.1.2 Questioning the Content/Vehicle Distinction

Claiming that vehicle properties can influence experiential content is not typically going far enough for the classical extensionalist, however. Extensionalists often take the claim further by arguing that, when it comes to the duration of the specious present, there is nothing to say about *content* at all, because the temporal "shape" or "structure" of the vehicle just *is* the duration of present experience. That duration is a span that may *include* content but is *not itself* content. If this is right, the analogies of the previous section are no longer apt, because there is no explanatory role for representation as far as temporal phenomenology is concerned.

Before continuing, it is worth pausing to fully defuse the common complaint that content/vehicle confusions infect this way of thinking, as so often happens in discussions about the relationship between brain and experience. As Rick Grush puts it: "One cannot just assume without argument, or without even addressing the issue, either that (i) properties of contents carried by vehicles can be read off any of the physical properties of the vehicles; or that (ii) relations between contents carried by sets of vehicles can be read off relations between physical properties of the vehicles," (Grush, 2006, p. 440).[17] It may be that the amount of time a given brain process takes is completely unrelated to the amount of *experiential* time elapsed. It remains logically possible that an instantaneous brain/informational state could represent a felt experience as a temporally extended process in the manner classical retentionalists have proposed. The same goes for the ordering of brain processes, which need not correlate to the order of events in experience. The latter helps retentionalists deal with illusions of event order in ways that extensionalism arguably cannot.

However, as much as it might be a mistake to think that vehicle properties necessarily correlate with content properties, it is just as much a mistake to think they necessarily *don't* correlate with content properties. Our previous examples of hieroglyphics and trees help to demonstrate situations where we would be wrong to ignore the vehicle properties and insist on an arbitrary connection between the representations and their contents, with the properties of the former irrelevant to the latter. Sometimes, the properties of the vehicles *do matter*; as the examples above have shown, it is often the case that a representation has its content in virtue of the properties of the vehicle bearing the content. As Grush goes on to say:

[17] Here Grush is specifically targeting Varela's (1999, 2000) attempts to naturalize Husserl's retentionalism by way of neural oscillation patterns, which Varela takes to be the neural correlates of time-consciousness, along with other similar attempts by Lloyd (2002, 2004) and Van Gelder (1996), rather than attacking extensionalism, which Grush disputes but is not directly concerned with in this passage. Nonetheless, the point is generally instructive.

> The fact that content/vehicle confusions are confusions does not entail that *all* properties of the vehicles of a representation are irrelevant for explaining the content carried by that representation. […] Some properties of the vehicle can be explanatorily relevant to the content they carry. Indeed, it would be hard to imagine a case where *all* vehicle properties were irrelevant. (Grush, 2006, p. 441)

Whether or not there is a strong connection between vehicle and content properties in the temporal case might then seem to be an empirical question, not an *a priori* one. In fact, Grush says that, while he agrees "to a first approximation" with Husserl's phenomenological analysis, he also thinks that "features of the neural information processing machinery in the central nervous system *are* relevant to those representation structures that underwrite the temporal aspects of our conscious experience," (ibid.). Indeed, it would seem absurd to claim otherwise without rejecting biology as the basis for experience altogether. This connection is also decidedly non-arbitrary. For our purposes though the takeaway is just that, in fact, vehicle properties can and often are relevant to content properties, whether directly (as with the isomorphism of hieroglyphics) or indirectly, through a yet-to-be specified relationship that the model must flesh out in a way that makes it clear why we should think such a connection would hold.

We can now return to bigger complaints about the content/vehicle distinction. An extensionalist might want to say the mistake is more profound than erroneously concluding that the absence of a necessary explanatory connection implies a necessary absence of explanatory connection. One might even think the entire content/vehicle framework is fundamentally misguided because temporal experience at its most minimal (as pure succession, say) cannot be grounded in the contents of representations if temporally structured experience is a precondition for subjectivity as we know it (see Chap. 3) and, consequently, also a precondition for representational content in the first place. We can see this when we consider that apprehension of content cannot take place in a single instant, but is a process that unfolds over time, necessitating a subjective succession. We therefore have the cart before the horse when using representational language to account for something like experiential succession which is, in reality, presupposed by it (NB: "experiential succession" not "experience *as of* succession").

From this perspective, intentionality assumes what it is recruited to explain, namely, subjective experience. Consciousness, extended in time, provides the enabling conditions for intentionality, not vice versa. We need some sort of prior conditions, what I have called temporal structure, which can be recapitulated in an ongoing fashion while an organism is conscious for this intentionality to obtain.[18] To illustrate further: the representing or intending of experiential objects like (let us suppose) "times" or "apples" is a cognitive *process*, and in order for these representations to exist as such there must be a temporally structured subject to represent something *to*. So, while intentionality might remain a key feature for

[18] John Searle, with similar antipathy to representational explanations, called such preconditions "capacities" and has claimed that "representation presupposes a nonrepresentational Background of capacities," (Searle, 1992, p. 178; see Chap. 8 for arguments to this effect).

phenomenology, the demand for some even more basic enabling scaffolding provides a further motivation for the extensionalist component of the theory that will be advocated here.

One might respond that representational content can exist in a subject-independent way, and the explanatory potency of the content/vehicle distinction can be maintained in the absence of subjectivity. The objection might be raised that symbols like *words* are not subjects and yet have representational content, so it cannot be that intentional content requires subjectivity. However, this argument rests on the flawed premise that such representations have content in the absence of subjects. In fact, representations like words really only have representational content in a derivative sense—otherwise they amount to nothing more than marks on a page. While this position is perhaps somewhat controversial, it may be enough for our purposes to remark that, even by the intentionalists' own lights, intentionality is generally held to be the mark of the mental and *only* the mental (this being Brentano's thesis, the central tenet of intentionalism; see Brentano, 1874, pp. 88–89).

Without experiential succession there would also be no subject because having a point of view on the world is essentially an activity, rather than an instantaneous state. Without this point of view there is consequently no standpoint from which content can be experienced. The idea of experiential contents without a subject is of course meaningless to speak of and so cannot in itself have explanatory purchase on subjectivity. Thinking of succession *fundamentally* in terms of contents will therefore not help us to understand succession. This is of course not to say representations have no explanatory value at all. The door is even still open for content to play a role in the explication of the feeling *as of* succession, which may be parasitic on an actual procession of successive content. The assumption though on the part of those looking to explicate experiential succession in terms of content and vehicles seems to be that a representational perspective will aid in this explication. However, such approaches bottom out attempting to explain minimal features of subjectivity like succession. We end up in a circular attempt to ground experienced succession in concepts that are incoherent without the prior existence of experiential succession.

The structure of experience, therefore, is really *the only place* to look for the fundamental properties determining the character of our minimal temporal phenomenology, such as the duration of the specious present or experiential succession. The preceding may seem to speak strongly (perhaps even too strongly) in favor of some form of extensionalism. However, separate problems stemming from the naivete of that view will shortly (in Sects. 6.3 and 6.4) lead us to consider adding content back into the picture to account for certain temporal phenomena.

6.2 Relationism, Resemblance, and the Jamesian Doctrine

Notwithstanding the benefits so far attributed to the extensional perspective, it is sometimes alleged that extensionalism is at a disadvantage relative to retentionalism if it claims that *successive experiences furnish us with experiences of succession*.

I will argue here that, while extensionalism *is* insufficient to account for all the richness of our temporal phenomenology, this insufficiency is *not* a consequence of the falsity of the claim that successive experiences furnish us with experiences of succession. Indeed, the proposal advocated here leans on extensionalism specifically to account for minimal features of temporal phenomenology—in particular, succession and the duration of the specious present. In part this appeal is made because extensionalism provides a much more straightforward explanation. Nonetheless, it might be thought that extensionalism is *too naïve*. Before further pursuing the advantages of a pluralistic approach then, we must first answer this common complaint against extensionalism, which should look far less significant upon inspection, especially when we recognize that *both* the unfolding of experience over time and intentional content can play an explanatory role for temporal phenomenology without contradiction.

Since the time of William James, it has been commonplace in the literature to invoke the slogan that "a succession of feelings, in and of itself, is not a feeling of succession," (James, 1890, p. 629). For brevity's sake, we can call this the "Jamesian Doctrine" (JD). JD is often seen as self-evident. However, in considering some of the examples of the last section, it might now not seem so obvious that there needs to be a disconnect between properties of contents and vehicles in the way demanded by retentionalists. We might think not only that vehicle properties are relevant, but *crucially* relevant, in that they can determine the temporal character of the resultant experience.

Christoph Hoerl (2013) has notably challenged the received dogma encapsulated in James' slogan. Hoerl finds the slogan repeated throughout the history of philosophy following James (e.g., Strong, 1896, p. 153; Husserl, 1917, p. 12; Paton, 1929, p. 318; Sellars, 1968, p. 232; Miller, 1984, p. 109; Dainton, 2008, p. 623) with widespread uncritical acceptance, particularly as grounds for accepting intentionalism[19] (with retentionalism being the preeminent form of intentionalism about temporal experience) and rejecting extensionalism. Hoerl notes however that, "it is not even clear why extensionalism and intentionalism should necessarily be seen as two distinct views of temporal experience, since an intentionalist might also maintain, as part of her theory, that episodes of perceptual experience must, as a matter of fact, be extended through a period of clock time," (Hoerl, 2013, pp. 375–376, n. 3). Indeed, something like the latter view would be sensible for an empirically minded retentionalist that understands the brain does not process and integrate information instantaneously.[20]

The real bone of contention between extensionalism and retentionalism seems to be about what is doing the work of *explaining* the phenomenology. More specifically, Hoerl sees the conflict as between a "representational view" of experience to

---

[19] Hoerl uses the term "intentionalist" to emphasize a commitment to the idea that intentional content determines phenomenal character, which is the default view among retentionalists. Intentionalism is a much broader thesis than retentionalism, however, and frequently appears in other domains of the philosophy of perception as well.

[20] Such retentionalists may include the "neurophenomenologists" mentioned in Sect. 6.2.

which retentionalism is committed, and a "relational view" of experience to which extensionalism is committed (Hoerl, 2013, p. 375). For an extensionalist adopting a relational view of experience, we should jettison talk of representations, which is why it can be misleading to say that extensionalists think vehicle properties determine content properties (given they reject the distinction in the first place). Rather, a relational view holds that experience consists only of "the actual items experienced and an entirely generic relation in which the subject stands to [that which is experienced]," (Hoerl, 2013, p. 379). Objects of experience[21] are therefore, at least on the latter view, *not represented*, but rather mere objects to which we stand in a particular *experiential* kind of relation.[22] To paraphrase Charles Travis, an advocate of a relational view of experience, we do not take the objects of our experience to be anything (i.e., they are not represented as being any particular way) before we have the experience; rather, we simply have the experience and *then* we are able take the experiential objects to be such-and-such a way (Travis, 2004, p. 65). Likewise, the relational extensionalist, "conceive[s] of the perceptual relation as something that itself unfolds over time" and it is this fact that results in experiential succession (Hoerl, 2013, p. 400). We should therefore not think we experience things *as* successive first and foremost; rather, we simply experience things successively.

Hoerl's contention that extensionalism really ought to be construed as "relational" is particularly significant in that it takes issue with how extensionalism has frequently been portrayed by intentionalists, namely, as a "resemblance" view. A resemblance view of extensionalism is one that holds "experience represents temporal features of the world by itself possessing those features," (Hoerl, 2013, p. 381). The previous section offered some initial reasons for why we might think vehicle properties might, at least in part, determine the character of experience, but in claiming this kind of "inheritance" of vehicle properties on the part of the experiential content, this way of thinking also presupposes some kind of representational structure. For a relational view of extensionalism, unlike a resemblance view, the previously discussed hieroglyphic and tree ring analogies are not apt. In those examples there is no relation of the right sort between a subject and the object of experience, despite there being a resemblance between properties of the representational vehicles and the content. The right sort of relation only comes about when an experiencing subject like us enters the picture.

Thus, on the relational view, talk of "vehicle properties" influencing "content" would be misleading because on such a view there are no representations and therefore neither vehicles nor contents. Instead, on the relational view there is just a subject and objects to which the subject is related. It is only when we think of extensionalism as offering a resemblance-based view that we fall into a

[21] I do not want to commit here to any particular view about the ontological status of objects of experience in saying this, because I think it is beside the point. I am not so much interested in what these objects are but in the fact that we stand in ever-changing relations to *something*, which is whatever our predictions determine is most probably the case.

[22] See, e.g., Campbell (2002, 2009), Martin (2002), Brewer (2004), Travis (2004), and Soteriou (2010), for examples and further discussion of the relational view in general.

representationalist mode of speaking and insist upon vehicles and contents and a resemblance between the two. However, extensionalism need not insist upon this at all if extensionalism is understood relationally instead of as committed to the idea that our temporal phenomenology *resembles* or inherits clock time.

One of the main reasons a relational view might be preferable is that, as I have mentioned, mental representation presupposes subjectivity, which should already be temporally structured. If this is right, then intentionality cannot be fundamentally explanatory, and we need to try something else for this domain. Instead, I would argue it may be better to think of the temporal point of view as consisting of successive relations between a subject and the objects of experience rather than thinking a series of intentional contents enable a temporal point of view to arise. If we adopt the former position (relationism), then the Jamesian Doctrine looks much less compelling than it does if we assume the latter position (maintaining the content/vehicle distinction). Relationism on its own it would rather trivially necessitate experiential succession, if experience is said to ultimately consist in a dynamic flux of relations to experiential objects. I would argue this is precisely what contemporary predictive approaches in cognitive science suggest and this is also the way that ERA analyzes experiential succession. How this works should become clearer in the next chapter when the successive relations between the predictive cognitive information processing system and the objects that the system takes to be actual (as the real causes of the stimuli entering the system) are more fully explicated.

Nonetheless, this is not to say that experience does not or cannot involve intentional content. Leaving intentionality entirely out of our description would seem to overlook the fact that some of our thoughts are *about* temporally indexed events. However, the fact that experiences are *about* things is insufficient by itself to explain how they come to be the experiences that they are. ERA can thus escape worries about the fundamental inadequacy of intentionalism as an explanation for minimal temporal phenomenology and at the same time can provide a non-circular grounding for experienced succession and the specious present by endorsing relationism to account for succession, while adopting a resemblance view to account for the duration of the specious present.

It is still the case that the actual temporal extension of brain processes and their resulting experiences at least provide a bound on the specious present and resolve the outstanding problems of infinitude and nesting from earlier in the chapter. Because these two explanatory strategies, relationism and the inheritance view, have different explanatory targets, they are not in conflict. They are both kinds of extensionalism in that they are non-intentional and take seriously real features of the world, specifically real features of the neural physiology, in accounting for temporal phenomenology, which, as extensionalism suggests, is temporally extended just as the processes themselves are (though we need not expect the relationship to be a 1:1 matching). More precisely, according to the view proposed here, the changing relations that account for experiential succession are themselves diachronically unified, as extensionalism suggests, by the real integration of ever-updating expectations over a period of time in the cognitive system, which specifies the duration of the finite specious present, however flexible and task- or modality-dependent this may be.

With these details in mind, the Jamesian Doctrine should seem less threatening to extensionalism than is usually imagined. We can now see the contention that "a succession of feelings, in and of itself, is not a feeling of succession" assumes an intentionalist perspective on experience, arguably begging the question against the extensionalist. Specifically, JD begs the question by assuming our experience cannot immediately inherit the temporal structure of the worldly basis for the experience, thereby rejecting a resemblance-based extensionalism out of hand. Less defensibly, the doctrine also dismisses the possibility that the feeling of succession simply *is* the successively changing relations in which we stand to various objects of experience, as certain naïve realists might hold. In this way JD also begs the question against any kind of relational extensionalism. As we have just seen, these questions are precisely what is at stake between extensionalists and intentionalists, so we cannot uncritically accept JD, as this would be tantamount to assuming intentionalism. Refusing to accept the Jamesian Doctrine, as I have suggested, removes a commonly emphasized obstacle to serious consideration of extensionalism. This is important because extensionalism really forms the foundation of the view presented here, in that ERA appeals to relational extensionalism to account for succession and appeals to resemblance extensionalism to account for duration.

6.3 Simultaneity Problems and the Pluralistic Strategy

ERA can also help to overcome two problems facing the individual analyses it incorporates. The first is a problem for extensionalism that Dainton has called the "Extensional Simultaneity Problem," or ESP (Dainton, 2017). The second is an analogous problem for retentionalism: the "Retentional Simultaneity Problem" or RSP (ibid.). Both problems, as their names suggest, are concerned with difficulties accounting for how there can be a specious present (as explicated by each view respectively) without the events experienced within it being experienced simultaneously. As we shall see, while the simultaneity problems present serious challenges to both extensionalism and retentionalism individually, a pluralistic explanatory strategy can help us escape these worries by addressing the problem for each with solutions from the other.

6.3.1 The Extensional Simultaneity Problem

The Extensional Simultaneity Problem refers to the question of how events that are "(i) experienced together, and (ii) experienced as present" can be "experienced as anything other than simultaneous," (Dainton, 2017). In other words, because extensionalism holds that experiences are spread over time, it is unclear how they could be unified in experience (i.e., co-conscious), yet *not* be experienced as simultaneous. In more technical parlance, this is the question of how so-called "diachronic" unity, or unity across or over time, is possible without collapsing into simultaneous experience. By endorsing diachronic unity in this way, extensionalists thus reject what Izchak Miller has called the "Principle of Simultaneous Awareness," (Miller, 1984, p. 109), which is the contention that in order for us to have a unified experience, all the contents of experience must be co-conscious at the same time, that is, they must be simultaneous, otherwise we would not be able to access them in our present experience. Retentionalists for their part tend to embrace this principle.

To illustrate ESP more vividly, we can consider an example of ordinary auditory perception. According to extensionalism, the experience of hearing the lyrics of the Rick Astley song "Never Gonna Give You Up… [etc.]" in sequential order is an experience that is extended in time (by my reckoning about 1.5 seconds). Part of this duration will comprise the specious present, the span that we take to be "now," perhaps roughly 200 ms. For the sake of this example let's say this involves the experience of the words "Never Gonna". Within the specious present that includes these auditory phenomena, "Never" and "Gonna" are both supposed to be experienced as if they are happening now and are thus in some way "co-conscious" to us in the sense of both being present experiences. However, it is indisputable that the two words occur in sequence, not simultaneously. If we grant that "Never" and "Gonna" are both present to us in some sense and the specious present is defined as a duration of actual experience, it looks like we are forced to conclude the two words would be heard as simultaneous. However, because it seems plain the two words are *not* experienced simultaneously, extensionalism looks as if it must be false, and there must be some other way of accounting for the specious present.

Extensionalists for their part may simply reject the intuition behind the PSA as an unjustified dogma that begs the question against them, just as William Stern did in 1897.[23] Stern even made the rejection of the PSA into its own "principle", namely that "mental events that play themselves out within a stretch of time can under circumstances form a unified and complex act of consciousness regardless of the non-simultaneity of individual parts," (Stern, 1897, p. 315). On this way of thinking, there is nothing about diachronic co-consciousness that is any more mysterious than synchronic co-consciousness, so the PSA provides no additional reason to think that the duration of the specious present must be explained by appeal to simultaneous contents *as of* an extended interval.

While there seems to be a commonsense intuition behind the PSA, i.e., that something about diachronic co-consciousness is *prima facie* less commonsensical than synchronic co-consciousness, extensionalists are right to be concerned that the reasons for thinking this is the case are not terribly compelling outside of a gut feeling.[24] Indeed, it is fully consistent with ERA to reject the principle as a dogma. As

[23] Thank you to an anonymous reviewer for encouraging me to include the extensionalist perspective here. See Dainton (2023, Sect. 5.1) for more commentary on this approach.

[24] If one is antecedently committed to presentism about objective time then it does look like *that* could be a compelling reason in itself to endorse the PSA, but here there are no such metaphysical commitments. It is possible, I think, that some of the intuitions in favor of the PSA actually come from a similar place as the supposed intuitions in favor presentism. This may be related to some-

Miller (1984) and others have defined it, the PSA is a commitment to the necessity of simultaneous "contents", which would seem to presuppose a general intentionalist framework towards experience in general that ERA denies. Denying the PSA is beneficial because, as Dainton puts it, "experienced contents which appear to be successive can really be successive, rather than existing simultaneously (as with retentional models), similarly episodes of consciousness which seem to persist through intervals of ordinary objective time can really do so," (Dainton, 2023). ERA, in its commitment to these more straightforward ways of accounting for succession and duration, thus also benefits from, and quite possibly requires, the rejection of the principle of simultaneous awareness.

However, it is also the case that our phenomenology of the specious present is not *just* a bare duration but also is colored by how events within it occur in relation to each other temporally, in the way the perception of a musical note is altered by what is expected to come after and what has come before. It is not *just* that the events are not experienced as simultaneous (which perhaps extensionalism, by rejecting PSA as a dogma, can sidestep) but that they are experienced as qualitatively differing depending on their temporal ordering.

ERA can appeal to retentionalism to resolve these issues: intentions allow for the contents of the diachronically unified experience to be experienced as not simultaneous but instead to be experienced however the intention specifies. Protentions and retentions allow for intentional content to be felt as if they were fading into the past or arriving from the future. These protentions or retentions *themselves* do not necessarily have to be simultaneous in order to have the intentional character they do but could be distributed over an interval of time. This interval would be important, on an inheritance view, in determining, directly or indirectly, the finite and variable duration of the specious present. Beyond this, the kinds of intentions that we have should influence the temporal quality of the experiential events and where they are experienced as standing in relation to each other.

Why should these intentions be distributed over an extended interval of real time? One answer is that the extension makes better sense of the specious present, as suggested earlier. We can solve the infinite regress problem facing retentionalism by maintaining the duration of the specious present depends on the temporal distribution of intentions. This is an explanatory appeal that is not open to a classical retentionalist, as it grounds the duration of the specious present in non-intentional properties. Strict intentionalists/retentionalists, in contrast, seemingly must commit to an arbitrary stopping point to the nesting intentions to avoid the regress, as previously mentioned. We shall see in the next chapter however that the predictive

thing of a commonly held "presentist bias", but I am not convinced it is justified by our experience, which even from a naïve perspective may lead us to endorse a non-instantaneous view of the present (and if language usage is any guide, we do often treat "now" as longer than a mere instant anyway).

cognitive apparatus commits us to temporally distributed intentions that together should determine the character of experience, so the substrate is already there if we endorse a predictive approach to the mind. A further *a priori* reason to make an explanatory appeal to the temporal distribution of intentions is to avoid a similar problem that faces retentionalism, and which concerns succession.

6.3.2 The Retentional Simultaneity Problem

The Retentional Simultaneity Problem refers to the question of how the contents of intentions can seem *successive* when they are actually simultaneous. As Dainton asks: "How is it possible for experienced succession to be generated by contents that exist only in the momentary present?" (Dainton, 2017). To illuminate the problem further we can return to the auditory experience considered previously. For a retentionalist, unlike an extensionalist, the intentions whose contents define our specious present are not necessarily extended over time, or if they are, the extension is explanatorily irrelevant. This means the "Never" and "Gonna" of "Never Gonna Give You Up" may be *represented* in the mind at the same time, but the contents of the representations are temporally indexed in relation to each other. The question is, how can simultaneous intentions come to have content that is experienced as if it is not simultaneous at all, barring mere stipulation that this is the case.

Retentionalists generally appeal to intentional content to solve RSP. The intentional content of retentions is meant to bestow upon them a unique phenomenal character, which, when combined with other intentions, results in a feeling of succession and not simultaneity. Yes, the intentions might all occur at the same clock time, but their contents are all *about* different times, hence they are not experienced as simultaneous but rather as successive. While tidy, the move here comes worryingly close to solving the problem by decree. We may still ask why appealing to "aboutness" leaves us any more enlightened than the original pre-theoretic intuition that we simply experience "never" and "gonna" as non-simultaneous.

Extensionalism, which does not posit intentions and embraces our pre-theoretic intuitions, does not arouse similar suspicions. On an extensional model experience is spread through time and, supposedly, can simply *be* successive and immediately perceived as such. The extensionalist typically enlists diachronic unity to do what intentionality was meant to do for the retentionalist. This is to say, the extensionalist claims there is a unification of events over time within one span of consciousness, which seems to be necessary for us to experience things as successive. However, invoking diachronic unity potentially runs back into ESP (if we find the PSA convincing), while by invoking intentions, retentionalism runs into RSP. Neither appears more promising than the other in relation to its respective problem. We seem to have reached an impasse.

Now we might think successive *content*, in contrast to a succession "in and of itself," is somehow better placed to explain our experience of succession just in virtue of it being content. Even so, it seems the skeptically inclined could reiterate the problem of succession for contents, as Dainton does, in the following way: "successive *content*, in and of itself, does not an experience of succession make [emphasis added]" (Dainton, 2010, pp. 107–108). Where the Jamesian Doctrine questioned the relevance of an actual succession to the character of experience, here the problem is shifted one level up, questioning the purported explanatory purchase of successive content on the character of experience. Similar to the original formulation of JD, it seems that, while we cannot *rule out* that successive content can make an experience of succession, we also cannot just assume that it does.

To answer the skeptic, perhaps the retentionalist can claim that content just *is* the stuff of experience, so there is no problem. However, such an answer amounts to relabeling the assertion that we have successive experience as the claim that we have successive experiential contents, providing no better explanation of how or why we come to have such a successive experience than extensionalism provides. Another reply might be for the retentionalist to claim that experience of succession is due to *further* content, i.e., higher-order content. However, the skeptic's reformulation of the Jamesian Doctrine can now be repeated for the higher-order content as well, and so on *ad infinitum*. Retentionalism then seems left without advantages over extensionalism in its treatment of succession.

6.3.3 The ERA Solution

Extensionalism avoids the Retentional Simultaneity Problem by maintaining experience is itself temporally extended but arguably runs into the Extensional Simultaneity Problem, namely that it is unclear why the temporal extension should not be experienced all at once if it is diachronically unified. While rejecting the PSA is an option for the extensionalist, it could be argued that merely doing this would still leave us with phenomenally underspecified events within the specious present. I therefore propose adopting intentions, not to account for the duration of the specious present or bare experiential succession, but to account for the differing quality of events dependent on their temporal relations with each other. I further propose adopting *temporally distributed* intentions to bypass RSP with the real temporal properties of this distribution having some bearing on the resultant phenomenology of the duration of the specious present. At the same time, the relations between the subject and what is actual are continuously changing because we construct the present through the ongoing generation of expectations about the world. This flux is the basis for experiential succession.[25]

According to ERA, intentions themselves are successively replaced and renewed, which is in keeping with an information processing picture (described more fully in the next chapter), whereby the system's expectations of the causes of incoming stimuli are continuously updated.[26] These continually updated expectations also exhibit what could be described as temporal content, in that they can be forward-directed and backward-directed, as in Husserl's scheme, meaning they differ qualitatively, though it is not their content that accounts for the experience of succession. There may also be higher-order representations with content to the effect that our experience is successive, for instance the content of the belief that experience is successive. However, fundamentally, experience just *is* successive in that expectations (usually understood or at least described as having intentional content) are successively replaced by new ones *and* this successive replacement is extremely relevant to our temporal phenomenology. As these expectations are updated, the relations between the subject and what the system takes to be actual is continuously changing. This is a generic relationship determined by the fact of their being expectations and the function that these perform for the entire multilevel hierarchy, namely, predicting the distal world through interaction with it and in doing so making that world subjectively real to us.

These expectations will still have intentional content[27] even as the generic relations they underwrite are continually succeeding one another just in virtue of continual replacement. In other words, the flux of expectations means there are ever changing relations to what is actual, but the intentional content of those expectations determines exactly what it is we take to be actual—the fine-grained phenomenal character or what-it's-like-ness of the perceptual experience. The kind of

[25] As previously mentioned, whether this is the basis for the "sense" of experiential succession or not is somewhat of an open question. As mentioned in Chaps. 2 and 3, in certain exotic situations, e.g., deep meditation and hallucinogenic drug use, people sometimes report losing the sense of succession, but it is doubtful that their experience is strictly frozen in time. If it were, the subject would not be able to concurrently go through the mental process of reflecting on the experience being frozen.

[26] This is not to suggest a strictly serial processing system operating on these intentions at a constant rate. As Wiese (2017) has suggested, there might instead be a complex, massively parallel, nested hierarchy of intentions together responsible for the salient aspects of the sensorium. The exact properties of this computational or other kind of architecture should be largely an empirical question.

[27] It seems convenient and somewhat standard to speak of intentional content when we speak of expectations. However, if it is possible to give a non-intentional/non-representational account of how expectations work within predictive processing, so much the better from my perspective. The important thing for ERA and its cognitive implementation, TIER (see next chapter), is ultimately the predictions themselves and how they are generated and not whether they are understood as intentional or not. See Kirchoff and Robertson (2018) for an example of a non-representational enactivist understanding of prediction that is ultimately grounded in covariance.

expectation (e.g., the timescale it concerns and whether it is past-directed, future directed, etc.) and how it relates to others in time also determines temporal properties of the event, but the content of the expectation (its intentional content) *per se* does not. Again, there is no doubt that higher order temporal representation can and probably is also occurring, but I do not think this should be invoked at an explanatorily low level on pain of the circularity discussed in Sect. 6.1.

All this is not to say that the doctrine that "succession, in and of itself, is not an experience of succession" is mistaken. A succession of numbers, for instance, is not an experience of succession. But in this case—the case of our minds—successively generated experiences are a succession of experiences and might thereby plausibly felt as such, even if it not necessarily so. This is related to Barry Dainton's contention that a special feature of so-called experiential content is that it does not exhibit an awareness-content structure, but instead, "become[s] conscious simply by coming into existence" (Dainton, 2000, p. 57). A succession of things coming into existence with *that* property is no mere succession, but an experiential succession. Provocatively stated, a succession of experiences, in contrast to a mere succession, *does* an experience of succession make.

What I am claiming is that a real structural feature, namely succession, is crucially relevant to our experience being experienced as successive. I am also claiming that experiential succession is fundamentally *not content* and should be further explicated with reference to structural features of the mechanisms underlying experience. These features are not necessarily *all* that is relevant to a full-blooded experience of succession, and ERA can allow for derivative, higher-order content as needed. It could be that in order to really "sense" the succession *as* successive or understand it as such, some degree of further reflexive awareness may be necessary, but this is a question that exceeds the scope of the current investigation into minimal temporal phenomenology. We may speculate, however, that some of the exotic cases described earlier in this book (see Chap. 2) involve a breakdown in this reflexive awareness, even if it is not possible for there to be a failure to generate successive experience without losing consciousness altogether.

In summary, we can now see that ERA cuts against the orthodoxy in multiple ways. First, it holds that what have been called "vehicle" properties, but which may be better termed "structural features," matter to experience. Secondly, I am claiming that a real succession, in this case of expectations, which continually alters the subjects' relations to what is real and what is not, does result in an experience of succession, bearing in mind the subtle point that whether this is an experience *as of* succession might be a further matter. The first claim is apparently vulnerable to a charge of content/vehicle confusion from intentional theorists, but, as discussed earlier, whether vehicle properties are explanatorily relevant is a contingent matter that not only cannot be ruled out, but, in the case of certain temporal properties like succession, seems quite plausible. The second claim is apparently vulnerable to that oft-repeated refrain about succession not giving rise to a feeling of succession. However, here too we can admit that, while it is true a succession by itself does not *necessarily* give rise to a feeling of succession, in the case of experience it seems plausible that a succession of experiences contingently would.

6.3.4 The Problem of Repeated Contents

A pluralistic explanatory strategy like ERA can also head off an objection sometimes made against specious present views, namely, the Problem of Repeated Contents (Mabbott, 1951). For extensionalists, the problem is that, if we posit overlapping specious presents to account for continuity, as many prominent versions of the view do (see Chap. 4, Sect. 4.3.2), we should expect to experience a repetition of equally vivid contents for the span of the diachronically unified specious present. This is a consequence of overlapping "acts of awareness" that each take in a certain interval but must overlap for them to be experienced continuously and connectedly (Dainton, 2010, p. 108). According to the Problem of Repeated Contents, the repetition involved in this picture is unfaithful to our actual phenomenology, which does not include such an over-proliferation of contents persisting and overlapping through the span of the present. To illustrate this problem further we can apply it to the example from the previous section. If one specious present includes the lyrics "Never Gonna" and the next includes the lyrics "Give You", they will be joined on an overlap model by a specious present including the words "Gonna Give". It would seem, however, that such an overlap would lead to a duplication of contents so that "Gonna" and "Give" are repeated in a way that they are actually not in our experience.[28]

Bertrand Russell, in his version of an extensionalist overlap model, claims that there will be no repeated contents because the overlapping parts ("Gonna" and "Give") are not actually repeated but rather shared: the parts of each specious present are not duplicates but identical. John Foster (1979), in similar fashion, thinks the overlapping specious presents share parts but do not repeat them. Going somewhat further in his own objection to the Problem of Repeated Contents, Barry Dainton denies the premise of the awareness-content distinction that the problem presupposes (Dainton, 2000, p. 57). If the extensionalist rejects the idea that the subject is aware of contents (instead of just having experiences), then it no longer makes sense to say that there are repeated contents on the overlap model, because this allegation assumes a model of perception in general that the extensionalist denies from the get-go. Instead, on the overlap model, there are only shared experiential parts that are diachronically unified.

Dainton's response reveals that question-begging commitments to the awareness-content distinction plague a lot of thinking on this topic. I agree that there is little reason for an extensionalist to adopt this distinction, and the extensionalist elements of ERA consequently do not. There simply is no awareness-content distinction when it comes to succession or the duration of the specious present. However, insofar as ERA still takes the way things are represented to us as an important contributor to our overall temporal phenomenology, it seems we still have to deal with experiential content at some point, even if we mean here that this content is what we experience as being the case based on our best probabilistic expectations.

[28] See Phillips (2010) and Pelczar (2010a, 2010b) for further discussion.

To put a fine point on it: ERA claims that experiential succession and the duration of the specious present are explained by structural properties of the way those intentions are temporally organized, rather than (exclusively) by their contents. In saying this, ERA rejects the PSA and the awareness-act framework just as Dainton does. We do not need to posit diachronically unified, temporally extended *contents*. Instead, there are only intentions spread over time, which *themselves* have perceptual contents. These contents are not *as of* durations or *as of* experiential succession in the minimal sense, and these contents are not responsible for our phenomenology of the same. While these contents qualitatively differ owing to the kind of intention they are, i.e., retention or protention, at no point do we face the problem of having the exact same kind of intentional contents repeated (unless that is how we happen to take the world to be).[29]

Also, like Dainton (2000; 2004), ERA takes the dynamic phenomenal flow of experience quite seriously. However, unlike extensionalist overlap views, which tend to appeal to a resemblance or inheritance picture to account for the entirety of temporal phenomenology, ERA only makes this kind of appeal about the duration of the specious present. To account for the phenomenon of flow, ERA takes a relational view, rooted in the continuous updating of our expectations about the world and indebted to a view of the mind as fundamentally a predictive engine. This is not a general relational view of perception but a relational view specifically with regard to experiential succession.

Why should we want a relational account of succession instead of an overlap model, when an overlap model seems capable of resisting the Problem of Repeated Contents on its own? One major consideration is that there does not seem to be anything inherent to overlapping specious presents that would result in a phenomenology of dynamism, just as there does not seem to be anything inherently dynamic about a snapshot view. But there *is* something inherently dynamic about ever-changing relations to what is actual, and this relational flux should result in experiential succession. The question of why the relations should be ever-changing is a different one to whether the dynamism is inherent to the model, but I would suggest the answer to that question is ultimately rooted in our evolving predictions we make about the world, which themselves are necessitated by the pressures to adapt and survive in the particular physical conditions we find ourselves in (discussed in greater detail in the next chapter).

On the retentionalist side, there is an analogous challenge to the problem of repeated contents or too much complexity. Here the idea is that, with all the retentions, protentions, etc., we should actually find that our experiential objects are multiplied to an unrealistic degree, somewhat like an echo or reverberation, but across all modalities. In our running example, we might imagine that the retentions of the lyrics "Never Gonna" should continue to be present for us as the song moves on, but they don't seem to be, at least not in anything like a vivid way. This would be what

[29] For further discussion of this problem, specifically as it relates to Broad's sense-data theory, see Dainton (2000, p. 145) and Gallagher (2003).

Dainton calls the problem of "lingering contents" (Dainton, 2000, p. 156). The answer from retentionalists is fairly straightforward however: this charge of repeated contents makes the mistake of treating intentional contents as if they are all intended in the same way and, in particular, of thinking that all intentions are meant to be experienced in the same vivid way, just as if they were merely at a temporal remove from each other. Precisely because they are *not*, the retentions, in the case of the example, do not feel as vivid as primal impressions, but perhaps more like a vague leftover impression of something having happened, fading into obscurity. In my view, it may be better and more literally accurate with the regard to the cognitive situation to say that they *inform* the current perception, but do not *linger* as the objection contends. I think this is a fair response if we believe the kind of intention matters. Events can take on different temporal qualities depending on the kind of intention carrying the content; for instance, protentional content is anticipated but is not a vivid part of our immediate experience.

Of course, it can be argued that we still have an overabundance of, e.g., retentions *qua* retentions, or protentions *qua* protentions. This is what Dainton calls the "clogging problem" (Dainton, 2000, p. 157). This is even more of a problem if combined with the lack of boundary a classical retentional view might have without an appeal to physical constraints. While establishing a boundary based on physical constraints can eliminate infinite complexity, we still might think there an unrealistic degree of phenomenal clutter. Similar to the previous response, one might think that just as protentions are understood as anticipations and not full-blown perceptions, something similar can be said about retentions, so it would not really be the case that retentionalism entails a literal clogging of, say, lyrical echoes. We do not ordinarily think having many anticipations about the future trajectory of the objects of our perceptions amounts to unrealistic phenomenal clutter, in part because these anticipations are not given to us in the same way as the most vivid actual percepts.

Still, Husserl maintains that entire "states of awareness" are retained with all of their own nested intentional structure, which sounds on the face of it to be quite demanding and perhaps phenomenologically inaccurate. Ultimately, I think these sorts of objections lose a lot of force when the relevant intentions (e.g., retentions and protentions) are reframed in light of their supposed cognitive basis within a predictive model. If a retention is really a retained prediction that influences the current trajectory of the predictive system as a whole, this no longer seems like we should expect phenomenal clutter, but rather a kind of insight that colors our perception with the knowledge of what has come before and the best guesses of what might come next. If this is right, then the clogging problem is not as serious as it seems, at least for the kind of non-classical retentionalism that I have in mind here.

Nonetheless, on the cognitive level, a very similar objection can be marshaled against hierarchical predictive processing models. The charge here is that there might be an over-proliferation of content associated with such a deep and complex nesting of predictions as we find in such models. The key to dealing with challenges of this type is to recognize that predictions are probabilistic, and no single prediction is terribly important on its own. Retentions and protentions in themselves thus do not necessarily constitute what is actual for the subject but what has been and could be actual to varying degrees, and no single prediction generates an experience but rather the experience depends on the aggregate of the hierarchical system. This is perhaps analogous to Husserl's contention that there is an invariant structure to consciousness such that the internal components cannot be meaningfully separated. To that extent only are those contents part of our experience, but they are no less an important part of it. This is just to say, expectations that are not assigned high probabilities at any given time will not be constitutive of perception but may nonetheless inform our perception. We thus should not expect a repeat of equally vivid experiences at a time or over time on these models.

With these objections out of the way, it may nonetheless be thought that ERA has too many moving parts and so falls prey to its own kind of "excessive complexity", in that it's needlessly duplicating work by appealing to multiple frameworks simultaneously to account for the same phenomena.[30] However, this objection mischaracterizes ERA as a traditional hybrid view that would try to explain all aspects of our temporal phenomenology through a single combination model. Rather, it is important to reiterate that ERA is actually a pluralistic explanatory strategy premised on the idea that we should expect very different aspects of phenomenology to demand different sorts of explanations, and as such sees virtue in ecumenism and vice in monomania. A heuristic that favors simplicity in these explanations should be understood to be nothing more than a heuristic, and one that is not appropriate to the domain of inquiry, which exhibits exceptional complexity.[31] Though it may be that this kind of pluralistic explanatory strategy is more complex than a simple overlap model, it is not complex for arbitrary reasons. It is instead complex because the mind and its cognitive machinery are themselves highly complex.[32]

While the core of the view may be seen as extensionalist in spirit, ERA still endorses the retentionalists' primary contention that the contents of intentions are not given in the same way, as retentions differ qualitatively in relation to each other as they slip further and further back. Maintaining intentional content also prevents events experienced within the span of a diachronically unified specious present from being experienced as simultaneous, avoiding the Extensional Simultaneity Problem as it might apply to temporally related events. While an extensionalist can get around ESP by rejecting the PSA and hold on to the notion that diachronic

---

[30] I thank an anonymous reviewer for pointing out this possible objection.

[31] See Chirimuuta (2024) for a more comprehensive resistance to simplifying assumptions towards hyper-complex systems like the brain. Following Chirimuuta, we can informally define the relevant sort of complexity here as a function of the "number and heterogeneity" of the components of a system at a time and across time along with the number and heterogeneity of the interactions between those components, along with the degree of "organizational depth", or functional nesting (wheels within wheels, so to speak), that we find in the system (Chirimuuta, 2024, pp. 9–10). I will just note here that fully appreciating this complexity would seem to entail that the analyses I'm proposing *also* involve unrealistic simplifications, which I do not doubt at all, however I see these and the general strategy and perspective I'm advocating as a potential advancement over what has come before.

[32] I speculate that recognizing complexity in biology seems to be easier than recognizing it in the phenomenology that depends on biology, but I think that neither should be presumed to be simple.

co-consciousness is fully possible, this in itself does not tell us why events are presented to us as having the order that they do relative to each other within the specious present. We might think, however, that properties of intentions are better placed to give the events of the diachronically unified span the particular temporal character they exhibit at different times.

Still, it is the real, rather than represented, flux of these contents, which rapidly succeed one another and are not simultaneous at any given time, which generates experiential succession. In fact, a phrase like "the experience *of succession*" is somewhat of a misnomer in that it implies succession is experiential content. Instead, on the view sketched here, we should rather say that experiences *are* successive, but we do not have experiences *of* succession at all, except perhaps in some higher-order sense. On this relationist picture, we should say we stand in successively differing relations to what is actual, and this is what is meant by saying experiences are successive.

Besides this critically important successive structure of experience, we also have experiences of events in temporal relations with each other as determined by intentional properties and a specious present whose boundaries are ultimately defined by the temporal extension of the processes realizing the intentions. These three elements comprise ERA's core commitments: succession as a physiologically necessitated flux of relations to the world and its objects, duration as an inheritance of temporal properties from physiological processes, and the temporal character of events and their temporal relations to each other as determined by the kinds of intentions (expectations) that ultimately generate what philosophers typically call experiential content.

6.4 Temporal Illusions

ERA exhibits certain advantages over the classical models in the way it can account for certain aspects of temporal phenomenology as well. Specifically, by appealing to retentionalism to account for event order, ERA can offset disadvantages faced by classical extensionalism, which on the naïve realist formulation struggles with cases of temporal illusions. While extensionalism has the virtue of parsimony, it does not as obviously have the virtue of easily accommodating our occasionally bizarre experiences. ERA lays claim to both virtues.

At first glance it may seem that, if we think features of temporal phenomenology (e.g., experiences of change, succession, etc.) are real features of experience,[33] then extensionalism seems to have the advantage of accounting for these features in a more straightforward, intuitive way. Change, for instance, would be experienced

[33] Dainton (2017) has called the view that such features are real in the sense that they are perceived directly "phenomeno-temporal realism" or "PT-realism." It is a view that is shared across both retentional and extensional models.

immediately, simply in virtue of there being changes over the course of the extended interval of experience, and likewise for succession and continuity. Indeed, if we inherit the processual character of the world, then questions that it makes sense to ask of experience flip upside down: whereas on a snapshot or retentionalist view the theory endeavors to account for why it seems to us like experience is dynamic, an extensionalist can more naturally say these features are fundamental—it is only when there is *unusual stability* in the process that we demand an explanation. We might think such "atomistic" positions as the snapshot view or certain forms of retentionalism involve unnatural, atemporal abstractions that stray from the reality of the dynamic experiential processes they aim to describe.[34]

Retentionalists, meanwhile, appeal to intentionality to account for features of temporal phenomenology, e.g., the second hand of a clock is represented *as* changing and thus it is experienced as such. For the retentionalists, this appears to have the unfortunate effect that changes are not necessarily felt directly, as they happen. We do not, on the retentionalist view, experience the changing position of the second hand *per se* but only experience the second hand *as* changing at a particular moment. Although this may seem like a disadvantage for retentionalism, we should also remember that extensionalism struggles to account for why real changes like the changing position of a second hand should have the additional felt quality of "changing" rather than just being experienced as occupying different positions in space, as with the hour hand (a problem identified by C.D. Broad, 1923, p. 351). A naïve form of extensionalism might claim that such an additional representational quality is just unnecessary when the property of changing is there to be picked up, in the object and its relation to its past and future states, all of which the subject directly perceives over an interval of clock time. If this view is correct, then extensionalism provides a more parsimonious explanation of the phenomena than does retentionalism.

Unfortunately, the naïve extensional view encounters a difficulty when it comes to the myriad temporal illusions that together indicate a less direct correspondence between the felt duration of experience and the actual duration of the events that are perceived. Retentionalism, by contrast, handles such discrepancies rather well. Intentions, by their nature, are also capable of *misdirecting* or *misrepresenting*, and this provides an intentional theory like retentionalism a ready-made solution to the problems of temporal illusions such as the flash-lag illusion encountered earlier. Experiential time dilation and contraction (that is, lengthening and shortening of experienced intervals in comparison to measured clock time) can also be explained by contents represented *as* occurring slower or faster or *as* occupying smaller or greater intervals of time. Extensionalists, so long as they insist upon a naïve, direct correspondence between felt intervals of time and actual intervals of time, lack this straightforward accounting of temporal illusions.

[34] I will not pursue this line of thinking in great depth here but see Nicholson and Dupré (2018) for an anthology arguing for the fundamental processual character of the world itself, and, in particular, biology.

An extensionalist *can* claim that the lag between the stimuli and our eventual illusory experience is great enough that whatever kind of strangeness results in the illusion occurs in this buffer zone or "pre-processing" (Dainton, 2017). However, this implies we are at more of a temporal distance from the things we perceive than it appears we really are, especially considering the speed with which we are capable of perceiving and acting. Barring epiphenomenalism (the view that conscious perception is causally inert), it is difficult to see how we could participate in fast-moving ball games or video games if we were systematically handicapped by the 80–100 ms lag that Dainton (2017) suggests extensionalists might endorse. Also, as some psychologists have suggested, long-lasting postdictive effects can extend over "several hundreds of milliseconds or longer" and would therefore require an unrealistically long lag on the extensional view (Herzog et al., 2020). Indeed, even Dainton's original figure seems questionable when we consider that, in esports, controllers with input lags of only 10 ms are desirable while over 50 ms of lag is potentially noticeable and uncompetitive. For a wireless controller to be perceived as smoothly tracking gameplay, "data must be refreshed 30 times or more each second," i.e., every 33 ms or so (Torvmark & Schweber, 2002). For these reasons and others, the "appeal to lag" does not seem like an elegant solution.

In contrast, as we have seen with the flash-lag illusion, postdictive effects offer illustrative examples of how retentionalism *can* elegantly accommodate temporal illusions. A further example of such an accommodation is in handling the "phi phenomenon," also encountered earlier. In one version of this illusion, a subject is first presented with a flash (usually a colored circle) at the top left of a screen and subsequently presented with a second flash of light at the bottom right. Although this object hasn't moved, the subject nevertheless reports continuous motion of the circle from the top left to the bottom right. If the flashed circles are different colors there appears to be a point halfway along the path of the motion of the circle where the circle changes color (known as "color phi"; Kolers & von Grünau, 1976). Of course, no such event has taken place—yet nonetheless it is perceived.[35]

The retentionalist has more than one interpretation of the phi phenomenon available to them, but here is one: the retentionalist can claim that the initial experiential content might have included just the first flash, while later events (the subsequent flash) change the contents of (a new) specious present to include motion across the screen, which is the experience that is remembered when the subject is asked to report the experience. Extensionalists do not have such an easy time offering explanations for these illusions. This is because, as Dainton points out, for the extensionalist, "there is just *one* token experience corresponding to the A-flash," which leaves us unable to account for the phenomenology of motion encountered subsequent to the presentation of the second flash (Dainton, 2017). If extensionalism were correct, we should expect subjects to report seeing either a stationary flash and *then* motion

[35] See Bach (2019) for an interactive example: https://michaelbach.de/ot/col-colorPhi/index.html

or to report seeing the situation accurately, but instead subjects report the illusory phenomenology as described.

With ERA, new possibilities emerge. We might think that, not only are things represented to us *as*, for instance, changing, but *also* that this change is experienced directly. Including both of these features results in a richer and potentially more accurate phenomenological picture. In the case of the Phi Phenomenon, the illusion might involve an abnormal representation of event order, but the temporal extension of the present experience remains stable as a result of the actual duration of our experience over time, as extensionalism suggests.

Besides a richer accounting of phenomenology, why else might we endorse this more complex approach? As we have already seen, in postdictive effects, subsequent events appear to affect earlier events. If these illusions do involve misrepresentations or post-hoc alterations of event order, it is difficult to see how a direct, naïve form of extensionalism could convincingly handle such cases,[36] while a retentionalist could say the events are represented *as being* in a non-veridical order. ERA can follow the retentionalist explanation, but also has the option of contending that where the intentions stand in relation to each other might be altered (a shuffling of the order of retentions, for instance) and that the contents of these intentions might change in response to this illusory ordering. Thus, on the latter explanation, some structural features of the organization of the intentions may be relevant, as well as the content of those intentions. ERA thus opens up new possibilities that were shut off to the classical views, while maintaining the explanatory resources to accommodate illusions and temporal variability. At the same time, ERA retains many of the considerations that motivate the extensionalist view, without making illusions inexplicable.[37]

[36] An extensionalist might try to give a disjunctive explanation of postdictive effects where the illusory cases tamper with something like our memory of the event (and are thus categorically unlike our ordinary "veridical" experiences), but arguably this way of handling the illusions appears ad-hoc and apparently empirically refutable as it would claim the illusions are not actually perceptual but derived from failed recollection (see Dennett, 1991, p. 115–126 for critical discussion of memory-based explanations; see Soteriou, 2020 for a general overview of disjunctive explanations).

[37] One might wonder how deeply *ontologically* committed we might have to be to the entities, like intentions, invoked by the explanatory strategy outlined above. I think the answer, given our epistemic limitations, is that we should remain commitment phobic. The intentional language deployed here can be seen as a useful explanatory strategy rather than a rigid ontology of the mind. See Chirimuuta (2024, Chap. 6) for a defense of agnosticism concerning the ontology of intentionality; see also Sprevak (2013) concerning fictionalism about neural representations. Pragmatic approaches like these provide a fertile ground for more specific explanatory projects (like that undertaken here), which freely draw on intentional language without committing to the fundamental reality of intentionality *per se*.

6.5 Historical Precursors

Naturally, this is not the first time a view has been suggested that cuts across the contemporary taxonomy. However, there are few explicit recommendations of an extensional-retentional analysis in the literature. Not least this is because, prior to the emergence of the current classification scheme, philosophers would not have thought of themselves as offering a combination view at all, but rather a distinctive stand-alone model of time-consciousness. Arguably, Shadworth Hodgson (1898) and C. D. Broad (1923) were two earlier philosophers that saw the appeal of an analysis with affinities to ERA, though they each laid emphasis on different aspects, and neither are full-throated advocates. Thomas Sattig (2019) has recently cast Anton Marty (1916) as an early proponent of an extensional-retentional hybrid view, but this interpretation seems to rely on a different understanding of the essential features of extensionalism than is typical. While Marty's view is inspiring in its own right, particularly in the way it connects to metaphysics, I will hesitate to class it as extensionalist, though it certainly bears strong similarities to later retentional analyses. These early views are instructive in that they demonstrate it was not always assumed, as we often find in contemporary times, that extensionalism and retentionalism are opposed.

6.5.1 Hodgson

Though perhaps he would not have described it in such terms, Shadworth Hodgson, a contemporary of William James, can be seen as advocating an early variant of extensional-retentional analysis. In *The Metaphysic of Experience*, Hodgson contends that, "[i]t is implied by the term empirical, as applied to any portion of experience, that this portion has some time-duration; and from this it follows that the least possible empirical present moment is one, in which perception and memory (in the sense of simple retention) are indistinguishable from one another," (Hodgson, 1878, p. 60). Hodgson here seems to be endorsing a form of extensionalism when he claims that experience has an "empirical" duration and that within this duration we cannot distinguish past from present in the normal way. It is not perfectly clear what Hodgson means when he says this is an implication of the "term empirical," but it could be that he felt the observable physical world is not divisible into instantaneous moments and because of this we cannot have experiential instants either.[38] On this basis, then, Hodgson might have concluded that duration is an essential feature of

[38] Hodgson seems to suggest something like this in the *Philosophy of Reflection*, when he says, "Crudely and popularly we divide the course of time into past, present, and future; but, strictly speaking, there is no present; it is composed of past and future divided by an indivisible point or instant. That instant, or time-point, is the strict *present*. What we call, loosely, the present, is an empirical portion of the course of time," (Hodgson, 1878, p. 253).

any experience, and indeed, any physical process. If this is accurate, then on the face of it Hodgson's position seems intuitive—if we think that experiences result from brain processes, and brain processes do not involve an instantaneous "end point" that produces experiences, then it does seem likely that both the process and the resultant experience would naturally extend over time.

In addition to the above considerations in favor of extensionalism, Hodgson is also sympathetic to the notion that more is required to account for temporal phenomenology than the experience's extension over clock-time alone. As for James and his successors, examples of experiences whose character seems to change depending on what has gone before (e.g. musical notes) led Hodgson to the conviction that "retention of a past in a present moment, now has been shown to take its place among the ultimate facts of experience, being involved in the simplest cases of perception, for which in fact, it is but another name," (Hodgson, 1898, p. 78). This mention of retention has led Barry Dainton (2023) to suggest Hodgson was advocating a "hybrid model of temporal experience that incorporates elements of both Extensionalism (over short intervals) and Retentionalism (over large intervals)."

However, the notion of retention may be more integral to Hodgson's view than it would be if it only featured in perception over large intervals. On this more robust hybrid view, there are intentions like retentions that needn't be simultaneous but instead can be spread over the duration of the minimum of consciousness, and the duration of their spread may influence the duration of what we take to be present. In other words, both extensionalism and retentionalism are operative over short intervals. Although his views might have changed over time, Hodgson's *Philosophy of Reflection*, passages of which are cited by James (1890, p. 690), seems to support the above interpretation. In keeping with retentionalism, Hodgson states that "…we have, in any artificially isolated minimum of consciousness, the *rudiments* of the perception of former and latter in time, in the sub-feeling that grows fainter, and the sub-feeling that grows stronger, and the change between them…" (Hodgson, 1878, p. 251). However, in keeping with extensionalism Hodgson also says that "this minimum has a duration" (Hodgson, 1878, p. 253). Given what we have seen from Hodgson previously, it is likely this is meant as an *actual* duration as opposed to the appearance of duration in experience that a classical retentionalist might endorse.

Hodgson also held a similar view regarding "imagination," which can be seen as roughly corresponding to protentions in his scheme: "…it is clear that the minimum of feeling contains two portions—a sub-feeling that goes and a sub-feeling that comes. One is remembered, the other imagined. The limits of both are indefinite at the beginning and end of the minimum, and ready to melt into other minima, proceeding from other stimuli," (Hodgson, 1878, p. 253). However, he also warns us not to take the minimum itself to be an instant, with all of its parts (i.e., retentions and protentions—what Hodgson has been calling "sub-feelings") occurring simultaneously:

> …the minima of consciousness are not *data* as minima, nor does consciousness come to us in distinct minima originally; but we it is who reduce it to minima artificially. And the very distinction of a minimum, so made, cuts it off from the continuous stream which it is part

> of, treats it *statically*, and thereby involves the assumption that its sub-parts are simultaneous. But this simultaneity of sub-parts, being introduced by our method, must not be reckoned to the thing analysed [sic]. (Hodgson, 1878, p. 252)

Hodgson is quite clear that we make an error when we "talk about the *present* as if it was a *datum*, as if time came to us marked into present periods like a measuring tape," calling such thinking "an instance of the separatist fallacy" rooted in misleading habits of thinking (Hodgson, 1878, p. 254). Ultimately, Hodgson's actual view on the subject remains somewhat obscure, but at the very least it seems that his account represents an early attempt at a plausible hybrid extensional-retentional view, and, in particular, here we can see the seeds of an attempt to show intentions distributed over time in an explanatorily relevant manner for understanding the temporal character of experience.

6.5.2 Marty

Anton Marty (1916) has been latterly interpreted, by Thomas Sattig (2019), as a defender of an extensional-retentional hybrid view. According to Sattig:

> Marty's approach is extensionalist, in so far as he allows the content of an experience of change to represent a temporally extended sequence of object-states. He does not, however, join the straightforward extensionalists [...], because he denies that the token experience itself has a temporal extension that matches the extension of the sequence it represents. Instead, the token state is, according to Marty, temporally unextended. (Sattig, 2019, p. 164)

Sattig goes on to explain the ways in which Marty's account is retentionalist, quoting the following key passage from Marty's *Raum und Zeit*:

> The intuition (*Anschauung*) of constancy and change involves: 1. that a continuum of temporal positions and their fillers is presented to us; and 2. that a judging consciousness is connected with the presentation of each of those positions and their fillers, which currently judges the presented as present, and then, for a while, as past. (Marty, 1916, p. 210)

Marty's view indeed looks retentional and is clearly very strongly committed to the explanatory value of representational content for the phenomenal character of experience, and this in the robust sense of the contents of "judgment." However, Sattig's characterization of Marty as extensionalist is somewhat suspect. Just because Marty allows "content of an experience of change to represent a temporally extended sequence of object-states" does not make him an extensionalist, at least not as extensionalism is understood here (see Chap. 4). In fact, Marty's contention that the token state is "temporally unextended" would seem to disqualify him from extensionalism entirely. Representing temporally extended experiential content is however fully compatible with retentionalism, even if it is a slightly idiosyncratic form of the latter. Marty, as a disciple of Brentano, is still keen to offload the explanation for the phenomenology of change onto intentional content.

The interesting element of Marty's view from Sattig's perspective is not that it is (or is not, as I tend to think) an extensional-retentional analysis, but rather that it is concerned with representing "actuality-status" (Sattig, 2019, p. 166). So, for Marty, it is not just that events are represented by retentions as, e.g., "past" (as might be standard for retentionalism), but that object-states, like past and future time-slices of a fallen leaf, are presented as "non-actual" in experience (i.e., not obtaining). Judgments of "reality status" and "temporal order" are both relevant to temporal phenomenology for Marty, and this helps his view align with his presentist metaphysics.

Unfortunately, since Marty's account is really just a form of retentionalism (as retentionalism is characterized here), it suffers from many of the same problems. As Sattig notes, "contemporary extensionalists claim to offer the most straightforward account of the representation of temporally extended sequences by holding that token experiences are themselves temporally extended…" (Sattig, 2019, p. 167). Marty's view, in rejecting the real temporal extension of experience, does not allow for this kind of account. Instead, for Marty there are instantaneous experiential moments wherein we represent temporally extended sequences of "object-states." As Sattig puts it, "many find this sort of representational power mysterious" (ibid.). The mysteriousness of representational power is of course a general problem for retentionalism, as previously discussed. A further worry is that the representation of a sequence of object-states does not in itself seem to provide solid grounds for the phenomenology of the sense of flow (ibid., p. 168) because a mere sequence is not enough to furnish us with a "feeling" of flow, as has also already been pointed out several times in these pages. Finally, Marty's presentism gets him into trouble. For Marty, there are actually no past or future object-states in existence, but rather only experiential representations of their absences, which means change, succession, continuity, and all the other phenomenological desiderata having to do with time, turn out to be part and parcel of an enormous and systematic illusion (ibid.). On such a view, perhaps nothing we experience is *non-illusory* because what is experienced is already always non-existent at the time of experience, which would seem to be an unpalatable consequence of any perceptual theory. Still, the emphasis on actuality-status is extremely insightful, potentially aligning with the sort of relationism concerning succession that, as we shall see in the next chapter, can be modeled within the framework of the predictive mind.

6.5.3 Broad

C. D. Broad also advocated a kind of extensional-retentional analysis in his 1923 *Scientific Thought*, with a greater emphasis on the extensional aspect than is found in my own proposal. To set up Broad's theory we must first imagine that we have awareness at a single moment, but this awareness is *of* experiential contents that are themselves extended through time, such that we can be aware now, in one moment, of contents at a remove of, say, .1 and .2 seconds in the past. Broad does not think we can ever *really* isolate awareness to a single moment, because, strictly speaking, moments do not exist—experience is always an ongoing, continuous process, rather than confined to an instantaneous point. Thus, Broad thinks awareness, like its contents, is always spread over an interval. For this reason, for Broad, any given specious present refers to an extended interval of awareness that is directed at extended contents. In broad strokes (pun intended), there are similarities with the extensional-retentional analysis given here, namely, that the intentions are themselves extended and the contents that the intentions are directed at are also experienced as extended.

There are some problems with Broad's account, however. Similar to how retentionalism faces the problem of surplus content, Dainton points out that Broad's view faces the problem of repeated contents (Dainton, 2023). Each awareness phase is directed at an interval of contents, say, .2 ms of notes in a song. However, the next awareness phase should also be directed at *at least* a part of those contents, meaning that we are aware of some of the content of the song twice, and three times, and so on, as the awareness phases stack up. This should lead to a repeated experience of the notes in the song for at least part of the given interval, but this does not seem to be what happens in our experience.

Additionally, Broad's theory must account for how the extended intervals of awareness are connected to produce an immediate feeling of continuity if we want to remain true to the immediacy thesis. The notion that we have a succession of intervals of awareness invites the charge that such a theory incorrectly characterizes experience as disjoint, when introspection reveals it is not. One solution is to appeal to a higher-order awareness, which is to say, awareness of the different spans of awareness. Unfortunately, such a move leads to a regress problem for Broad's theory. In its unmodified form then, the obstacles facing Broad's view are numerous.

6.5.4 Comparison with ERA

The kind of extensional-retentional analysis on offer here (ERA) is more defined than its predecessors and will become further specified as the mechanism underpinning it at the cognitive level becomes clearer in the next chapter. However, at the level of phenomenology we can think of ERA as sharing certain features with Broad and Hodgson's earlier models. To summarize ERA's commitments, we can start by outlining a standard conception of the retentionalist model. On such a model, there are intentions—in this case, retentions and protentions, with the former directed at the past and the latter directed at the future. Given this commitment, the problem of constraining experience to a plausible specious present and resisting a potential infinitude of intentions and intentions of intentions then motivates the adoption of a kind of extensionalism.

According to ERA, intentions themselves do not exhibit infinite recursion. This I take as a brute fact of the lower-level cognitive implementation, which will be laid out in further detail in the following chapter. However, the intentions *are* distributed over clock time in a relevant way. The real temporal spread of the distributed intentions does some of the explanatory work that in classical retentionalism is performed by higher-order intentionality. Note that the intentional content here concerns worldly causes, not time or times, unless derivatively. There is thus a certain priority to actual extension—the fact that intentions are spread over time is the more fundamental temporal fact of the system.

Contrary to the position of strong intentionalism, this actual extension over time has a direct impact upon phenomenology in constraining the extension of our experience to the particular span the intentions are spread over. The duration of actual cognitive processing should matter in some way. At the same time, the ever-changing relations in which the subject stands to the objects of perception, as determined by the continual replacement of predictions about the world, accounts for experiential succession. But here too this flux of relations is diachronically unified as extensionalism would hold, albeit through cognitive integration of an entire multi-level predictive hierarchy over time.

Again, the motivations for and details of this position should become even clearer when the cognitive implementation is spelled out further in terms of predictive processing in the next chapter. Given that, in this scheme, retentions and protentions are only giving events temporal qualities like "earlier than", "later than", and "simultaneous with",[39] there is no danger of the Problem of an Infinite Present. Rather we are left with a picture that accords with a finite and measurable specious present. This is not to say that the specious present would not vary—indeed, for different tasks and different modalities there may be variance, as ultimately the extension of the intentional entities is underpinned by an extension of relevant physical processes in the brain over clock time, which itself varies, potentially in a task and modality dependent way. The contention here is just that the details of the extended processes underlying our experience are directly relevant to the experience itself.

6.6 Conclusion

A pluralistic explanatory strategy to account for widely differing phenomena is not only a viable option, but a sensible one. We might envision a model where intentions are distributed over time not just incidentally, but in a way that is explanatorily relevant to the felt duration of the experienced present. We might also think experiential succession results from ever-changing relations between a subject and the objects of experience. Thinking in these *extensional* ways helps avoid the problems retentionalism faces in accounting for the specious present as well as the retentional simultaneity problem, while also providing a straightforward account of duration and succession. At the same time, ERA maintains the relevance of temporally indexed content to temporal phenomenology. Thinking in this *retentional* way helps to avoid the extensional simultaneity problem while also providing a plausible accounting of temporal illusions. The ultimate upshot is that, if we avoid treating

[39] I.e., "B-ish" properties according to McTaggart's (1908) scheme.

retentionalism and extensionalism as necessarily opposed, we can begin to develop a more fruitful understanding of experience that synergizes with predictive approaches in cognitive science, as we shall see in the next chapter.

References

Bach, M. (2019). *Colour Phi Phenomenon*. 147 Optical Illusions & Visual Phenomena. Retrieved 23 February, 2022, from https://michaelbach.de/ot/col-colorPhi/

Brentano, F. (1874/2015). *Psychology from an empirical standpoint*. Routledge Classics.

Brewer, B. (2004). Realism and the nature of perceptual experience. *Philosophical Issues, 14*, 61–77.

Broad, C. D. (1923). *Scientific thought: A philosophical analysis of some of its fundamental concepts*. Kegan Paul.

Campbell, J. (2002). *Reference and consciousness*. Oxford University Press.

Campbell, J. (2009). Consciousness and reference. In B. P. McLaughlin, A. Beckermann, & S. Walter (Eds.), *The Oxford handbook of philosophy of mind* (pp. 648–662). Oxford University Press.

Chirimuuta, M. (2024). *The brain abstracted: Simplification in the history and philosophy of neuroscience*. MIT Press.

Dainton, B. (2000). *Stream of consciousness: Unity and continuity in conscious experience*. Routledge.

Dainton, B. (2004). The Self and the Phenomenal. *Ratio, 17*(4), 365–389. DOI: 10.1111/rati.2004.17.issue-4. https://doi.org/10.1111/j.1467-9329.2004.00261.x

Dainton, B. (2008). The experience of time and change. *Philosophy Compass, 3*, 619–638.

Dainton, B. (2010). *Time and space* (2nd ed.). Acumen.

Dainton, B. (2017). Temporal consciousness. In E. N. Zalta (Ed.), *Stanford Encyclopedia of Philosophy* (Fall 2017 Edition), https://plato.stanford.edu/archives/fall2017/entries/consciousness-temporal/

Dainton, B. (2023). "Temporal consciousness". In E. N. Zalta and U. Nodelman (Eds.), *The Stanford Encyclopedia of Philosophy* (Spring 2023 Edition). https://plato.stanford.edu/archives/spr2023/entries/consciousness-temporal/

Dennett, D. (1991). *Consciousness explained*. Little, Brown and Co.

Foster, J. (1979). In self-defence. In G. MacDonald (Ed.), *Perception and identity: Essays presented to A. J. Ayer with his replies to them* (pp. 161–185). Macmillan.

Gallagher, S. (2003). Sync-Ing the stream of experience: Time-consciousness in Broad, Husserl, and Dainton. *Psyche, 9* (10). Available online at Retrieved February 23, 2022, from https://journalpsyche.org/files/0xaabb.pdf

Grush, R. (2005). Internal models and the construction of time: Generalizing from state estimation to trajectory estimation to address temporal features of perception, including temporal illusions. *Journal of Neural Engineering, 2*(3), S209–S218.

Grush, R. (2006). How to, and how not to, Bridge computational cognitive neuroscience and Husserlian phenomenology of time consciousness. *Synthese, 153*, 417–450.

Herzog, M. H., Drissi-Daoudi, L., & Doerig, A. (2020). All in good time: Long-lasting Postdictive effects reveal discrete perception. *Trends in Cognitive Science, 24*(10), 826–837.

Hodgson, S. H. (1878). *The philosophy of reflection* (1st ed.). Longmans.

Hodgson, S. (1898). *Metaphysic of experience Vol. I*. Longmans.

Hoerl, C. (2013). A succession of feelings, in and of itself, is not a feeling of succession. *Mind, 122*(486), 373–417.

Hohwy, J. (2013). *The predictive mind*. Oxford University Press.

Hohwy, J., Paton, B., & Palmer, C. (2016). Distrusting the present. *Phenomenology and the Cognitive Sciences, 15*, 315–335. https://doi.org/10.1007/s11097-015-9439-6
Husserl, E. (1917/1964). In M. Heidegger (Ed.), *On the phenomenology of the consciousness of internal time* (1893–1917) (J.S. Churchill. Trans.). Indiana University Press.
James, W. (1890/1980). *The principles of psychology*. Henry Holt.
Kirchoff, M. D., & Robertson, I. (2018). Enactivism and predictive processing: A non-representational view. *Philosophical Explorations, 21*, 264–281.
Kolers, P. A., & von Grünau, M. (1976). Shape and color in apparent motion. *Vision Research, 16*(4), 329–335.
Lloyd, D. (2002). Functional MRI and the study of human consciousness. *Journal of Cognitive Neuroscience, 14*(6), 818–831.
Lloyd, D. (2004). *Radiant cool: A novel theory of consciousness*. MIT Press.
Mabbott, J. D. (1951). Our direct experience of time. *Mind, 60*, 153–167.
Martin, M. G. F. (2002). The transparency of experience. *Mind & Language, 4*, 376–425.
Marty, A. (1916). *Raum und Zeit [Space and Time]*. Niemeyer.
McTaggart, J. M. E. (1908). The unreality of time. *Mind, 17*(68), 457–474.
Miller, I. (1984). *Husserl, perception and temporal awareness*. MIT Press.
Nicholson, D. J., & Dupré, J. (Eds.). (2018). *Everything flows: Towards a Processual philosophy of biology*. Oxford University Press.
Northoff, G. (2016). Slow cortical potentials and 'inner time consciousness'—A neuro-phenomenal hypothesis about the 'width of present'. *International Journal of Psychophysiology, 103*, 174–184.
Northoff, G. (2023). *Neurowaves: Brain, time, and consciousness*. McGill-Queen's University Press.
Paton, H. J. (1929). Self-identity. *Mind, 38*, 312–329.
Pelczar, M. (2010a). Presentism, eternalism and phenomenal change. *Synthese, 176*(2), 275–290.
Pelczar, M. (2010b). Must an appearance of succession involve a succession of appearances. *Philosophy and Phenomenological Research, 81*(1), 49–63.
Phillips, I. (2010). Perceiving temporal properties. *European Journal of Philosophy, 18*(2), 176–202.
Phillips, I. (2014a). The temporal structure of experience. In V. Arstila & D. Lloyd (Eds.), *Subjective time: The philosophy, psychology, and neuroscience of temporality* (pp. 139–158). MIT Press.
Phillips, I. (2014b). Experience of and in time. *Philosophy Compass, 9*(2), 131–144.
Sattig, T. (2019). Experiencing change: Extensionalism, retentionalism, and Marty's hybrid account. In G. Bacigalupo & H. Leblanc (Eds.), *Anton Marty and contemporary philosophy* (pp. 153–174). Palgrave Macmillan.
Searle, J. (1992). *The rediscovery of the mind*. MIT Press.
Sellars, W. (1968). *Science and metaphysics: Variations on Kantian themes*. Routledge & Kegan Paul.
Soteriou, M. (2010). Perceiving events. *Philosophical Explorations, 13*(3), 223–241.
Soteriou, M. (2020). The disjunctive theory of perception. In E. N. Zalta (Ed.) *The Stanford Encyclopedia of Philosophy* [online encyclopedia]. Winter 2010 Edition. Retrieved February 23, 2022, from http://plato.stan-ford.edu/archives/win2010/entries/perception-disjunctive/
Sprevak, M. (2013). Fictionalism about neural representations. *The Monist, 96*, 539–560.
Stern, L. W. (1897/2005). Mental presence-time (Nicolas de Warren, Trans.). In B. Hopkins and S. Crowell (Eds.), *The new yearbook for phenomenology and phenomenological philosophy* (Vol. 5, pp. 310–351). Noesis Press.
Strong, C. A. (1896). Consciousness and time. *Psychological Review, 3*, 149–157.
Thompson, E. (2010). *Mind in life: Biology, phenomenology, and the sciences of mind*. Harvard University Press.
Torvmark, K. H., & B. Schweber. (2002). *EDN* [Electrical Design News], *47* (20), 30–36.
Travis, C. (2004). The silence of the senses. *Mind, 113*, 57–94.

Van Gelder, T. (1996). Wooden iron? Husserlian phenomenology meets cognitive science. *Electronic Journal of Analytic Philosophy, 4*. Accessible online at https://ejap.louisiana.edu/EJAP/1996.spring/vangelder.1996.spring.html

Varela, F. (1996). Neurophenomenology: A methodological remedy for the hard problem. *Journal of Consciousness Studies, 3*, 330–350.

Varela, F. (1999). Present-time consciousness. *Journal of Consciousness Studies, 6*(2–3), 111–140.

Varela, F. (2000). The specious present: A neurophenomenology of time consciousness. In J. Petitot, F. J. Varela, B. Pachoud, & J.-M. Roy (Eds.), *Naturalizing phenomenology: Issues in contemporary phenomenology and cognitive science* (pp. 266–314). Stanford University Press.

Wiese, W. (2017). Predictive processing and the phenomenology of time consciousness: A hierarchical extension of Rick Grush's trajectory estimation model. In T. Metzinger & W. Wiese (Eds.), *Philosophy and predictive processing* (Vol. 26). MIND Group. https://doi.org/10.15502/9783958573277

Chapter 7
TIER: Temporality as Iterative Expectation Revision

The previous chapter introduced a specific pluralistic way of accounting for fundamental aspects of temporal experience, which I have called ERA. This chapter will spell out how such an account might be implemented. Specifically, this chapter will explore how the advent[1] of predictive processing (or predictive coding[2]) approaches in the mind sciences can help answer the question of how the brain constructs the experiential present. Predictive approaches like those championed by Jakob Hohwy (2013) and Andy Clark (2016) exhibit certain features, namely an in-built temporal structure and dynamic, environmentally sensitive model update system, that make them ideal bedfellows with the phenomenological analysis outlined earlier. I will argue that seeing the mind as a predictive engine helps us understand how experience is constructed, and in doing so, shows us why and how we construct our present. In particular, I will argue that the process of expectation revision—the core of the predictive engine—is responsible for minimal temporal phenomenology, i.e., experiential succession and the extended dynamic present. According to this proposal, the temporality of experience at its most basic is not itself a product of a

[1] While Hermann von Helmholtz (1860) introduced the notion that perception is dependent on a mechanism of unconscious inference, more recently sophisticated variants of the general predictive paradigm have been championed by Karl Friston (2003, 2010), Jakob Hohwy (2013), Andy Clark (2016), Anil Seth (2021), Mark Solms (2021), Parr et al. (2022), along with very many others. See Hatfield (2002) and Swanson (2016) for more on the historical roots of predictive processing in Helmholtz.

[2] Attempts to understand the operation of the nervous system through the lens of probabilistic data compression go by the name "Predictive Coding" (Rao & Ballard, 1999). The broad notion that the brain is "Bayesian" or can be usefully described as such (Doya et al., 2007), the notion, known specifically as "predictive processing," that cognitive systems like ourselves seek to minimize prediction error (Clark, 2016), and the far-reaching research program of Active Inference and the Free Energy Principle which embeds predictive cognition in more general physical principles (Parr et al., 2022), can all be described as "predictive approaches" and share many commitments. Sometimes these labels are treated as equivalent despite differences.

C. A. McKenna, *Constructing the Present*, Studies in Brain and Mind 24,
https://doi.org/10.1007/978-3-031-89571-5_7

cognitive mechanism but instead a feature of the mechanism's ongoing operation. For brevity's sake we can call this proposal, which understands temporality as iterative expectation revision, "TIER."

For proponents like Jakob Hohwy, the predictive framework "explains not just that we perceive, but *how* we perceive: the idea applies directly to key aspects of the phenomenology of perception," (Hohwy, 2013, p. 1). Beyond simply *influencing* subjective phenomenology, by Hohwy's lights, "it is *only* this idea [predictive processing] that is needed to explain these aspects of perception," (ibid). One of the great virtues and challenges of predictive processing is that it is supposed to provide an account of how and why our subjective experience is the way it is, and even, according to some of its proponents, why there is anything it is like to be conscious at all.[3] Thus we ought to expect that predictive processing absolutely *must* have something to say about why our subjective temporal experience is the way it is. Fortunately, predictive processing not only can provide an account of temporal experience, but also is uniquely well positioned to do so.[4]

This chapter's goal is to answer the question of how the present is constructed by isolating the most significant aspects of the predictive mechanism for the minimal phenomenology of time-consciousness. As previously mentioned, this minimal phenomenology includes experiential succession and a privileged, dynamic, extended present for the experiencing subject. My proposal is relatively straightforward: the predictive mind's ongoing construction of the present is responsible for minimal temporal experience. The main argument for the proposal is also straightforward: if we grant the basic tenets of the predictive approach and grant that the mechanism has a purchase on phenomenal consciousness in general, then iterative expectation revision should provide the structural foundation for temporal phenomenology because within this framework expectations are ultimately responsible for the character of phenomenal consciousness, i.e., what experience is like for us as subjects. No further explanatory appeals (e.g., to higher-level priors, meta-cognition, etc.) need be made beyond the basic predictive mechanism in order to account for the aspects of time-consciousness that are most characteristic of subjectivity *per se*. These minimal features can be accounted for by appeal to structural features of an operating predictive mechanism rather than by appeal to the content of predictions themselves. Notably, these minimal features *do not* include higher-level awareness of the fact that we are experiencing things successively or the fact that we experience a privileged dynamic present.

---

[3] Not every advocate of the framework agrees that predictive processing can satisfactorily answer the so-called "Hard Problem of Consciousness"—i.e., the problem of why there is anything it is like to be conscious (Chalmers, 1995)—but Andy Clark at least has voiced cautious optimism (Clark, 2016, p. 239; 324 n. 26).

[4] Of late, more and more attention has been directed to the unique explanatory upshots of the predictive approach for time-consciousness as such. A non-exhaustive list of work to this effect would include Kiebel et al. (2008), Hohwy et al. (2016), Friston and Buzsáki (2016), Wiese (2017), Salti et al. (2019), Montemayor and Wittmann (2021), and Kent and Wittmann (2021), but the topic remains relatively understudied within the rather large and ever-growing interdisciplinary literature on the predictive mind.

In this vein, some recent predictive proposals (e.g., Hohwy et al., 2016; Montemayor & Wittmann, 2021) seem to exhibit a common thread which is captured in simpler terms by Hoerl and McCormack's (2019) proposed "temporal updating system," which is meant to be "phylogenetically and ontogenetically" primitive and responsible for an organism's most basic temporal phenomenology (as opposed to the more sophisticated temporal reasoning that adult humans can engage in). To account for the existence of temporality as a feature of experience at all, we need to look at the simpler system, and the predictive approach provides an opportunity to flesh out the idea that the need to update our models of the world is ultimately the origin of time-consciousness.

At a minimum, predictive processing accounts describe cognitive systems that are engaged in a constant process of revising predictions[5] about the environment they are embedded in. It is the content of the models that result from the updating of these predictions that constitute the perceptual world when the entire multilevel process is considered. Updating these predictions can therefore generate a continual replacement of experiential contents such that we experience time flowing. If this is right, we do not experience time *itself* (whatever that would mean), but rather our perceptions are structured temporally and constantly updated by the perceptual system, in conformity with the extensional-retentional phenomenological analysis (ERA) offered in the previous chapter.

With regard to the relationism about temporal succession that ERA proposes, we might see the constant revision and replacement of our expectations as a succession of ever-varying experiential relations to the objects of experience, with these relations and their objects being defined by the content and kinds of expectations the system generates concerning the causes of incoming stimuli. It is thus more accurate to say that we perceive or experience things *temporally* rather than that we perceive time. In this sense, TIER adheres to an overall adverbialism about temporal experience—time is not something we experience but instead *how* we construct experience.

At the same time, the real duration of cognitive processes as instantiated neurophysiologically should matter to the felt duration of experience, by which I mean, the specious present. For TIER and predictive approaches generally, experience is not "completed" at any one particular level or with any one particular prediction but arises out of the continual operation of the system as a whole, which integrates information over time. More specifically, the multilevel processing hierarchy that predictive approaches endorse is necessarily offset in real time, a fact that demands the integration of levels in this hierarchy also be diachronic. TIER sees this brute

[5] Often the term "prediction," "expectation," and "model" are used somewhat loosely and even interchangeably. These terms should not be taken literally. They are technical terms or theoretical posits that ultimately stand in for particular neural behavior, which can be usefully described as predictive. It is not as if the subject is literally predicting, expecting, and modeling, but rather that activities vaguely similar to these higher-level cognitive acts are occurring at a deep level in the system that is supposed to constitute the subject. See footnote 10 for the specifics on exactly how this picture looks computationally, without the potentially misleading terminological gloss.

physical fact about the operation of the cognitive system as relevant to the resultant temporal phenomenology, specifically in terms of the felt duration of the interval of the present, which is after all, according to predictive approaches, the ultimate upshot of the working of the system. The present is extended in the way that it is just *because* the cognitive integration of our predictions is diachronically unified.

Finally, there is a retentional component to ERA that is also captured by TIER. According to the predictive approach, it is the content[6] of our predictions in aggregate that determines the quality of experience. Endorsing the predictive view means a general view of experience like this, but TIER refrains from placing the quality of temporal experience at that level, instead casting it as a more fundamental feature of the way that the predictive system operates. Nonetheless, the predictive model is well-suited to provide a basis for a retentional-style view because of the nesting of predictions with targets at various timescales.

This chapter will begin by describing the predictive approach in greater detail and then address why, on this model, it makes sense to understand temporality as rooted in iterative expectation revision (TIER). It will then specify how our earlier phenomenological analysis (ERA) might be grounded in the information processing picture promoted by the predictive approach. My proposal here is meant to be suggestive but should provide reasons for thinking the predictive approach is a particularly fruitful research paradigm through which to investigate the cognitive implementation of the phenomenology of subjective time.

7.1 Predictive Processing: An Overview

Predictive processing (or "PP") is ultimately a theoretical framework that derives specific explanations for myriad mental phenomena from one overriding principle. That principle, stated succinctly, is that the brain is an "ever-active prediction engine," (Clark, 2016, p. 52). Our perception results from the working of this predictive engine. Jakob Hohwy calls this, "the organizing principle for brain function as such,"[7] and believes that, "it should then encompass and illuminate all aspects of perception and action…" (Hohwy, 2013, p. 101).

---

[6] The term "content" is a term of convenience as it seems like a natural way of characterizing that which is predicted. However, if predictions can be successfully recast in terms of covariance so that we dispense with representational language, so much the better. I prefer to remain as non-committal as possible here about the metaphysical status of predictions and their alleged "contents" because what I propose is not so much concerned with that as it is concerned with the fact that there exist something like predictions which target events at different times in different temporal ways (e.g., *as though they just occurred*) and bear temporal relations to each other, which affects how such predictions are experienced. Whether this situation actually requires a rich ontology of full-blown representation is not a question I wish to take up here. My honest opinion is that I am deeply skeptical of such an ontology but find it serves as a useful stand in for what may really be going on.

[7] Hohwy is speaking specifically about the ongoing process of prediction error minimization here.

More precisely, the claim is that brains like ours make Bayesian predictions that take into account sensory input to adjust the expected probabilities of the causes of stimuli in an iterative fashion, all with a view to adaptively guiding behavior—behavior which, in turn, furnishes us with yet more information with which to update the probabilities of our predictions. To make the best predictions, the brain must always be updating its predictions in a way that takes into account and adjusts for the discrepancy ("error") between those predictions and the sensory signal coming in, through a process known as "prediction error minimization" (PEM).

If this account is correct, we are not passively engaged in the absorption of environmental data but actively participating in the two-way street of perception by generating and adjusting predictions in light of incoming information. As Hohwy explains, it is crucial to prediction error minimization that the brain has, "access to not only the incoming sensory data but to a comparison between this data and expectations about what the data should be, under a model of the world. The difference between these two is prediction error, which in turn is then a measurable quantity for the brain, and something that can act as a feedback signal on the way its models of the world are chosen and their parameters revised," (Hohwy, 2013, p. 75). What we have in predictive processing is therefore a theory about the function of the brain and how that function is carried out on a mechanistic level. What we still do not have (yet) is a theory of the physical implementation of this functional mechanism, nor a theory describing or grounding phenomenology.

Although not strictly targeting these other levels of analysis, there are nonetheless inferable connections between them such that predictive processing appears to lead in the direction of a unified multilevel framework, even to the extent of providing something like "bridge laws" allowing for "translation" between levels of analysis. If predictive processing is correct, for instance, it should be the case that our subjective experiences are directly determined by the operation of this systematic computational mechanism—there could be no change in the phenomenology without a corresponding change at the computational level.[8] As Wiese and Metzinger explain:

> [Predictive processing] is not directly a theory about the underlying neural processes (it is computational, not neurophysiological), but there are more or less specific proposals of how predictive processing can be implemented in the brain (see, e.g., Engel et al., 2001; Friston, 2005; Wacongne et al., 2011; Bastos et al., 2012; Brodski et al., 2015). Moreover, it seems that at least some of the principles which can be applied to descriptions on subpersonal (e.g., computational or neurobiological) levels of analysis can also be applied to descrip-

[8] In other words, the phenomenology supervenes on a computational level predictive mechanism, which in turn supervenes on the neurobiological processes occurring at the physical level. Although not usually made explicit, proponents of predictive processing most naturally assume a physicalist, realist metaphysics (i.e. all the facts of the universe are physical facts and there is a mind-independent reality) along with a functionalist theory of mind allowing for multiple realizability (i.e., it is the predictive function of the brain that matters for consciousness and this function could in principle be realized by other physical material and still be conscious in the same way).

> tions on the personal level (e.g., to agentive phenomena, the structure of reasoning, or phenomenological reports which describe the contents of consciousness). (Wiese & Metzinger, 2017, p. 2)

The predictive processing framework, to its proponents, thus seems to promise theoretical unity, parsimony, elegance, interconnectedness, and so forth, which in part explains its recent rise to prominence, especially when few competitors can offer such fine-grained descriptions of disparate mental phenomena at various levels of analysis in such a conceptually consistent way. As Hohwy puts it, "the framework is extremely parsimonious, with a simple mechanism at its heart, replicated throughout the hierarchy and yet able to fulfill a number of computational functions," (Hohwy, 2013, p. 75). This is an attractive feature for a biologically instantiated information processing system to exhibit.

7.1.1 The Core Predictive Mechanism

Beyond the basic claim that the brain is a Bayesian prediction-making machine, predictive processing accounts almost always include several more specific features. The most essential features are the following, as identified by Wiese and Metzinger (2017):

1. Top-down processing
2. Statistical estimation
3. Hierarchical processing
4. Prediction
5. Prediction error minimization
6. Bayesian inference
7. Predictive control

Top-down processing (1) refers to how, unlike a more traditional functionalist view of mind, where sensory data only flows up from sense organs to a passive "seat of consciousness," in predictive approaches expectations of bottom-up signals also flow downwards. Top-down signals affect the lower levels and bottom-up signals, which ultimately derive from environmental and bodily stimuli, affect the higher levels. At every level there are **statistical estimations (2)** of the probability of signals occurring at the lower levels of a **multilevel hierarchy (3)**, with the higher levels responsible for increasingly complex signals and a set of very top levels underpinning conscious experiences. Each level can affect and be affected by all other levels to some degree, either directly (in the case of those immediately above, below, or adjacent) or indirectly (in the case of more distant levels).[9] The

[9] There is controversy concerning the degree of what is known as "cognitive penetrability," which refers to the ability of high-level cognition (e.g. beliefs) to affect perception, but it is generally held that no level or set of levels is functionally isolated from the others (see Hohwy, 2013, pp. 117–139).

levels here refer to an abstract conception of the functions of the cognitive system and its interconnections, somewhat like a wiring diagram, and should not be confused with a description of the physical instantiation of the cognitive processes.

The information generated by statistical estimations is used to **predict (4)** the incoming sensory signal, and the error between the predicted signal and the incoming signal is sent up the hierarchy to influence the higher-level model, which then adjusts to **minimize prediction error (5)**. This process of prediction error minimization (PEM) is supposed to "approximate exact **Bayesian inference**"[10] **(6)** (Wiese & Metzinger, 2017, p. 8). The exact details of how this works need not concern us greatly, as it is not the level upon which the present discussion will turn, but I have included a more detailed primer on Bayesian inference in the last footnote for those interested and the details may clarify how information from previous expectations is literally retained as posterior probabilities when there is a model update.

Finally, **predictive control (7)** maintains that an organism acts to change its sensory input to fit its predictions. This means that behavior, at least some of the time,[11] is a symptom of the general goal to reduce prediction error. Although the proximal target of any given prediction is only ever lower-level activity, the distal target is ultimately the causal structure of the world itself. In the end, if perception is to adaptively guide an organism's behavior through a potentially hostile

---

[10] Bayes' theorem is a mathematical rule determining the probability of a hypothesis given certain prior information. In Bayesian inference the goal is to compute the posterior probability of a certain hypothesis, i.e., a truth-evaluable statement about the world. It is called the *posterior* probability because we want to know the probability of the hypothesis *given* some previously ascertained data. The posterior probability can be formalized as p(hypothesis|data), where "|" means "given." It is computed using the following equation, known as Bayes' rule:

$$p(\text{hypothesis} \mid \text{data}) = \frac{p(\text{data} \mid \text{hypothesis}) \times p(\text{hypothesis})}{p(\text{data})}$$

In plain English, this means the probability of the truth of a hypothesis given some known data is equal to the probability of that data given the truth of the hypothesis (called the "likelihood") multiplied by the probability of the truth of the hypothesis on its own (called the "prior probability") all divided by the probability of the data occurring on its own (called "marginal likelihood," i.e., the probability of the "evidence"). In speaking of the mind, the entire numerator or just the specific likelihood p(signal|hidden causes) is often called the "generative model" (Wiese & Metzinger, 2017, p. 14). Predictive Processing holds that each level of the hierarchy implements Bayes' rule to adjust predictions based on incoming sensory information and increase precision in future predictions. To formalize Bayes' rule in terms familiar to perception, we can say that, "the posterior probability $P(X|Y)$ of the world state X given sensory input Y is proportional to the product of the likelihood $P(Y|X)$ and the prior probability $P(X)$," (Doya & Shin, 2007, p. 9). Generally, it is thought that probability *distributions*, as opposed to single-state estimates, are preferable in guiding action (ibid.). Posterior probabilities are also reused in iterative fashion as new priors for future predictions.

[11] For the more avid supporters of predictive approaches, all of the time. The degree to which this single mechanism accounts for an organism's behavior is a controversial subject that need not detain us here.

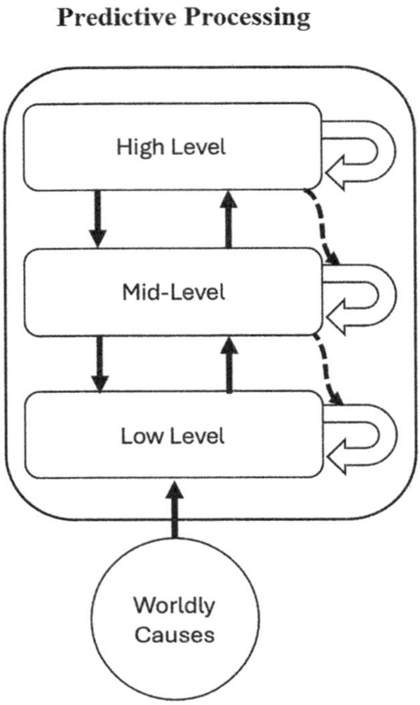

Fig. 7.1 Hohwy's predictive processing hierarchy (adapted from Hohwy 2013, p. 67). Predictions flow from the top down while error signals flow from the bottom up, with each adjusting the model at the level it feeds into. Additionally, each level exhibits self-inhibition given the expected precision of its model (circular grey arrows), and each level also feeds back predictions about expected precision. On the physical implementation level, Hohwy believes superficial pyramidal cells of the cortex may be responsible for feeding forward while deep pyramidal cells feed back (Hohwy, 2013, p. 68). Note that the terms "model," "expectations," and "prediction," are sometimes used interchangeably in the literature. These terms seem to invite a kind of misleading anthropomorphizing, but we should not see this process as being at all related to the high-level cognitive processes we explicitly engage in when we "expect," "predict," or "model" the world consciously. In fact, much of this framework describes subpersonal cognitive processing, below the level of our awareness

environment, the target of those predictions must be the dynamic world it inhabits (Fig. 7.1).

7.1.2 Predictive Processing and Temporal Structure

The most important feature of the predictive framework for temporal phenomenology is the constant Bayesian model updating occurring at each level in the hierarchy. When we consider how the predictive approach understands the relationship between the mechanism and consciousness, the connection to temporal phenomenology becomes clearer. Hohwy et al. (2016) explain where they see experience entering the picture:

> In creatures like us, it is natural to assume that perceptual inference maps on to *conscious* experience (Hohwy et al., 2008; Hohwy, 2013). If this assumption is granted, conscious experience at any time would be determined by the set of predictions, distributed over the hierarchy and mixing, or convolving, hidden causes in a certain way, that at that time is best at minimizing the current prediction error. This yields an overall *hypothesis* with the highest posterior about the current sensory causes, which is therefore used to generate predictions for the ensuing sensory input. Because this hypothesis encompasses multiple levels of the hierarchy it seems a good candidate for delivering the richness of conscious perceptual experience. (Hohwy et al., 2016, p. 319)

If the best overall hypothesis or expectation, rooted in a distributed set of predictions about the causes of sensory stimuli, is responsible for the phenomenal character of experience, then it must be that as the expectations are revised, the phenomenal character of experience is replaced by something else, even if that something else is nearly identical because the degree of revision was minimal. Hohwy, Paton, and Palmer recognize this consequence of the predictive engine when they say that "…as soon as one hypothesis wins and thereby determines perceptual experience, it will begin to fade away and another hypothesis takes its place," (Hohwy et al., 2016, p. 323).

As described in the previous section, according to predictive approaches there are expectations generated by a model (the "generative model"), incoming sensory signals, and errors produced by the interplay between signals and expectations at each level of the hierarchy. The posterior probability of a certain cause given certain signals is constantly updated based on new sensory information in what Andy Clark calls "a continuous process of sensory prediction in which the target is a kind of rolling present," (Clark, 2016, p. 18).

However, rather than *targeting* the present as such, I contend it is this process that *constructs* the present, understood as a privileged subjective temporal perspective. The predictive targets are in fact just the causes of the input. The common notion of the present, i.e., the notion of "present" which is related to a "past" and a "future" which lie behind and in front of the present along a linear dimension (cf. McTaggart, 1908), does not yet make sense applied to this context, as the system has not yet created the conditions for the spatialization of the concept of time. In other words, the conceptual present requires further metacognitive processing, but that which we will eventually come to think of as our present is already constructed at this basic level of processing.

Because the system implements Bayes' rule to arrive at posterior probabilities of the hidden causes of sensory data, the predictive view entails that the mechanism responsible for our perception has a built-in temporally indexed informational structure. As Clark further explains, "the percept itself [is] a prediction-driven construct that is always rooted in the past (systemic knowledge) and anticipating, at multiple temporal and spatial scales, the future," (Clark, 2016, p. 18). Aspects of the constructed present thus include information about the system-relative "past" and "future." This incorporation of prior and potentially later information into the subjective present provides a basis for the widely popular "specious present" accounts of temporal phenomenology that claim the experiential present is spread out over time in some way. According to predictive processing, however, such a temporal structure occurs at both sub-personal and personal levels of the hierarchy, remaining a feature of the system even in the absence of consciousness.

Our predictions are constantly updated in response to the incessant flow of sensory information, so even if the percept and the prediction remain practically the same, a computation exploiting incoming information is still occurring. In such a scenario, despite the apparently changeless world, nonetheless a duration elapses from the perspective of the system, because an ongoing series of updates still takes place (although this is not to say the system is "conscious" of any duration at this stage). It is this constant Bayesian update, even when the predicted causes of stimuli remain practically identical, that gives us our experience of time passing whenever we are experiencing anything.

As we have seen, in predictive processing, posterior probabilities concerning the hidden causes of incoming sensory information are necessary prerequisites for perception. The proposal here is just that the update of this posterior probability enables, but does not guarantee, the experience of time passing, or, more precisely, experiential succession. We end up with a situation where we as subjects, who come into being *qua* subjects through this evolving predictive process, stand in successively differing relations to objects of experience. As the updates occur, the same or differing experiential objects are successively related to us as being present (i.e., as existing) or not, with all the affordances for action that relationship might entail. The experience of the passage of time is thus not a percept itself but falls out of the *way* that all percepts are dynamically created. This is the core of the view advanced here, i.e., Temporality as Iterative Expectation Revision (TIER), according to which temporality is understood as an intrinsic feature of the operation of the cognitive mechanism.

The actual content of the experience thereby generated, insofar as it is informed by the past and anticipatory of the future, can also be said to have retained prior iterations in the current best models that ultimately ground experience. At the same time the system aims to predict the incoming future stimuli at multiple timescales and over multiple spatial dimensions, including predicting its own behavior, on micro and macro-scales. This may even include retrodiction. Given that earlier and later events reverberate throughout the predictive hierarchy in a rather complex nested fashion, the parallels with retentionalism should now be starting to come into view. What times and timescales the predictions are over should have an impact on how the content of the relevant predictions are experienced.

Beyond this, the continual updates of expectations within the hierarchical system should not be understood to be occurring instantaneously. By physical necessity, this would be impossible. Rather these updates are *processes* occurring over time and offset in time, yet diachronically integrated. The way the ongoing iterative expectation revision is temporally integrated across multiple levels in the hierarchy to form a holistic system grounding a subjective experience is unclear, but if it is indeed integrated, and if this system truly grounds experience, then the temporal properties of the window(s) of integration[12] of the predictive models should be expected to influence the finite experiential duration of the specious present. This is the span over which the flux of relations determined by successive model updates would be considered present by the subject that is created out of this active system. In other words, the resultant interval is the specious present. According to the predictive approach we should say this experiential window just *is* the subject, with this ongoing activity grounding the entirety of subjective existence.[13] So it is that I suggest the predictive mind constructs the present. In Sect. 7.2 I will go on to describe how the combination of the above features of iterative expectation revision yields a phenomenological picture that is best analyzed in the way described in the previous chapter, namely, the Extensional-Retentional Analysis (ERA). Before doing so, however, I will address the role that some common additional commitments might (or might not) play with respect to time-consciousness.

7.1.3 Additional Commitments

In addition to the core claims of predictive processing outlined above, there are several further features found in most prominent predictive processing accounts. Such ancillary commitments often include: the notion that attention is really the result of precision weighting of predictions, the notion that we can describe what the brain is doing in predictive processing as testing hypotheses, the notion that the ultimate goal of the system is to minimize "free energy," and the so-called "ideomotor principle," which maintains that action and perception are inextricably linked in such a way that perception is rooted in predictions about how the organism's actions affect the environment in ways that result in structural regularities that can act as perceptual feedback (Wiese & Metzinger, 2017). Furthermore, two camps have formed on either side of a debate about whether predictive processing entails "environmental seclusion," that is, whether the environment is merely inferred (a position known as internalism; see Hohwy, 2013) or whether the boundaries of the system extend out into the body and the world itself (a position known as externalism; see Clark, 2016).

For our purposes, we can remain neutral on most of these things. However, the notion of "precision estimates" is less controversial within the context of the predictive processing framework and may well play a role in modulating the temporal experience. While the system is ultimately engaged in inferring worldly causes of sensory input, it is also engaged in determining the level of perceptual noise generated by a given state of the world. Certain conditions are far "noisier" than others,

[12] These windows of integration are not the same as any particular sensory modality's window of integration, and the problem of accounting for how the system is unified will not be answered in strictly the same way as accounting for how information across and within modalities is integrated.

[13] If we don't say something like this, we may fall prey to the Cartesian Theater objection.

that is, there are conditions that inherently involve much greater uncertainty than others, and it is useful to know about these states. A dimly lit room for instance provides less reliable information about the objects within it than a brightly lit one—in the latter scenario it is quite possible to mistake a coat rack for a person because of the indeterminacy of the incoming data. To deal with this problem and allow the perceptual system to exhibit context-sensitivity, predictive processing accounts invoke self-inhibiting reflexive connections at each level of the hierarchy. Hohwy describes this structure in greater detail:

> […] Lateral connections within the same hierarchical level serve to decorrelate prediction units such that when a particular hypothesis begins to emerge as having the highest posterior probability other units are progressively prevented from influencing inference (Friston, 2002a, b). In an uncertain situation, such as when there is ambiguous context or noise, a number of different hypotheses can concurrently seek to predict the input.... However, when one hypothesis is deemed good enough the activities of other hypotheses should begin to recede because whatever they can successfully predict is most likely just noise—those hypotheses are *explained away*.... an evolving prediction error signal that can continually be explained away well by a particular hypothesis can be assumed to be reliable and should thus be weighted more in the message-passing economy.... This notion of weighting prediction error according to their [sic] reliability is central to how the system deals with noise and uncertainty, and in turn, to the nature of attention. (Hohwy, 2013, p. 61)

The system makes predictions about expected precisions, so besides just inferring causes of incoming sensory data, the brain is also constantly making predictions about the reliability of that data. This can be thought of as a "second-order perceptual inference," (Hohwy, 2013, p. 65). Perception is concerned with making the best hypotheses for the available data, so reliability is important. When incoming data is assigned more reliability, there is greater precision, that is, less variance in the data. The brain thus wants to maximize precision, at least in scenarios where it is very important to do so, and predictive processing contends that attention is the means by which this is done.[14] We generate predictions about expected precision and then adjust the amount of weight given to predictions about hidden causes based on the expected precision of the conditions encountered by the organism. An increase of precision weighting is quite literally the act of attending according to this framework.

What does precision weighting have to do with temporal experience? We imagine attention increasing visual acuity, auditory acuity, etc., as when you attend more closely to the lyrics of a song on the radio or the text on a street sign while driving. But attention has also been shown to modulate the experienced speed of temporal passage. Specifically, attention to a task has been shown to result in time dilation (Frey, 1990; Mattes & Ulrich, 1998). It could be the case that attention, through increasing precision weighting, allows the brain to make more and finer-grained Bayesian updates, which translates into finer temporal resolution. This increase in temporal resolution could be experienced as a slowing down of the passage of time

[14] Interestingly, William James long ago had a similar thought concerning the role of attention: "[T]he mind is at every stage a theatre of simultaneous possibilities. Consciousness consists in the comparison of these with each other, the selection of some, and the suppression of the rest by the reinforcing and inhibiting agency of attention (James, 1890, p. 288)."

for the percept because a greater amount of detail is being processed at a finer temporal grain than normal. Presumably the opposite scenario, i.e., time contraction/speeding up, would then obtain under conditions of minimal attention. This mechanism then may do some justice to the folk wisdom that "a watched pot never boils."

7.2 ERA and the Predictive Mind

It remains to be explained how the basic principle of temporality understood as iterative expectation revision (TIER) corresponds to the extensional-retentional phenomenological analysis (ERA) proposed earlier. To put a fine point on what has been previously discussed we can isolate the claims involved in TIER to the following:

1. The phenomenology of succession results from continual revision of expectations at all levels of a multilevel hierarchy engaged in predicting the causes of incoming stimuli. This is in keeping with a relational vision of extensionalism.
2. The operation of the predictive system occurs in real time and its actual temporal extension bounds the interval of the present (the specious present), albeit not necessarily directly. This is in keeping with a resemblance or inheritance view of extensionalism.
3. The character of temporal phenomenology is also determined by specific aspects of the information processing architecture, namely predictions and error signals, involved in the continual revision of expectations. This is in keeping with retentionalism (Fig. 7.2).

Under the predictive processing scheme, residual error signals used to influence future updates can be seen as analogous to retentions, while top-down predictions about what will happen next satisfy the future-directedness criterion for protentions. Husserl named the unity of "primal impression-retention-protention" that we experience the "living present" (Husserl, 1917, p. 434): the present of our experience, whose invariant structure, as Evan Thompson puts it, "does not move in or through time" and yet also "underlies any appearance of flow whatsoever, including the appearance of consciousness to itself as flow," (Thompson, 2010, p. 326). Husserl felt that positing an invariant temporal structure was necessary to give time its dynamic quality, because a series of static moments, even if those moments are extended in time, would lack an inherent dynamism. TIER captures this "living present" by appealing to the functional components of the predictive mechanism underpinning subjective experience. A few big caveats to this match-up with Husserl's view are in order, however.

On the view presented here, there isn't really a primal impression or *absolute now* for the subject, but rather only continuous Bayesian updates that occur throughout the hierarchy including elements (error and predictions) that can be usefully interpreted as retentional and protentional. There is no privileged "now-point" *for the whole system* within the interval of the specious present: there are only errors

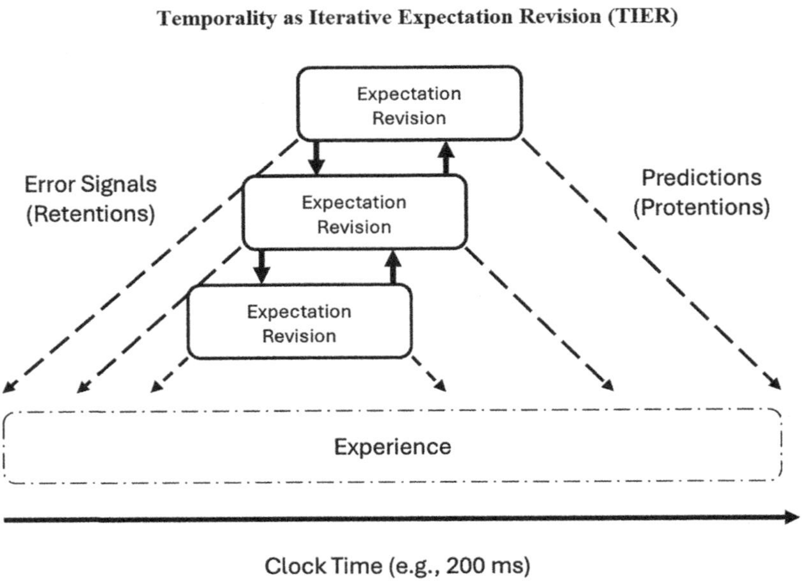

Fig. 7.2 Illustration of temporality as iterative expectation revision (TIER). This proposal is an implementation of the extensional-retentional analysis (ERA). Continuous Bayesian updates, or "expectation revisions," distributed over real time, provide indexical now-phases *for each level*, which together constitute the window of the specious present for the system as a whole. Error signals may play the role of retentions with future-oriented predictions acting as protentions. Only three levels of the multilevel hierarchy are shown, but distant levels of the hierarchy would be further offset in time with the total offset of variable but finite temporal extension. The dotted arrows represent the two-way interaction between levels common to hierarchical predictive approaches (see Sect. 7.1). This mechanism is of course a gross abstraction simplifying highly complex parallel brain processes. Note also that no particular level or single update is responsible for the whole experience. Exactly where and why subjectivity enters the predictive processing picture remains a "hard" question that is outside the scope of the current inquiry. However, if the predictive approach is indeed a more or less accurate description of the activities of the mind that generate experience, TIER may explain how such an experience comes to have a temporal structure simply in virtue of the operation of the predictive mechanism

and predictions, which contribute to the updates. In this way, there are some affinities to Dennett and Kinsbourne's influential Multiple Drafts Model (Dennett & Kinsbourne, 1992). Roughly, this is the idea that the brain entertains multiple competing models (drafts) at any given time and it is only upon introspection that one is selected as "the" experience we think we have had. In this context, we can think of there being "multiple nows" for the subject, rather than a single instant that is identifiably "now" for the whole system (see Dennett, 1991, Chap. 5). The greatest divergence between TIER and other predictive models of time-consciousness, which will be discussed in more detail in Sect. 7.4, is that the contents of the update (and error and predictions), i.e., what is estimated, are not intervals of time, but *probabilities of occurrences within time*. As a consequence, it is the updating itself, and not the contents of the updates that ultimately generates the phenomenology of a "living present."

In accordance with retentionalism, at each level of the hierarchy, updates are made possible by appeal to information about the past and predictions about the future. As Wiese and Metzinger put it, "the prior codes the information the agent already has; the likelihood codes how the domain about which the agent already has information is related to the domain of the new information obtained," (Wiese & Metzinger, 2017, p. 6). These are features of updates at a single level. Throughout the system, more robust traces of the past are present in the form of the error signals passed up the hierarchy. There is therefore not one single analogue for retentions in the model proposed here, but several.

ERA is a quasi-retentional analysis, albeit not necessarily a strongly representational one, and TIER accommodates this as well. Although there is almost as much disagreement over what representations are as whether and how they play a role in perception, it is generally accepted that representations must be about distal events and separable from the distal events they are about. Paweł Gładziejewski (2015, 2016) claims predictive processing is representational because "PP's generative models perform a representational function by acting as detachable information bearing structures which stand-in for the features of the environment," (helpfully paraphrased by Krzysztof Dolega, 2017, p. 2). On the other hand, in PP, predictions and errors need not represent but merely covary with proximal causal mediators, i.e., the immediate level below or above. If one insists the generative model *is* representational, on the proposed account the contents of that representation are not "time" or "times," but only objects, broadly construed, already structured *temporally*. The experience of time is a process found closer to the bone, so to speak, than a purely representational view allows for. Even with a commitment to representational content, we can still see time not as a kind of content in itself but as a consequence of the way that perceptual content comes to be.

If not really "about" times *per se*, retentional models may still be *intentional* in the most basic sense of having direction towards an object. Indeed, Husserl defines retentions, primal impressions, and protentions as each being "directed towards" different phases of time. This is to say there is a distinction in the temporal orientation of say, a retention of events past and a protention of future expected events. That said, maintaining intentionality is a relatively minimal ontological commitment compared to taking time itself or times to be a represented object. The idea is just that certain elements of the predictive system are directed towards the past or future, which doesn't amount to saying anything more than we already knew about error signals and predictions.

In Husserlian terms, priors and error signals can be seen as "traces" of the past, while predictions about the future are evidently predicted "traces" of the future—both of which are necessary to ensure that succession is felt *as* succession, rather than disconnected percepts with no temporal relation to each other that happen to occur one after the other. While the system is of course still making estimates, these estimates do not take intervals of time as their objects. Instead, such estimates ultimately only take worldly causes as objects. The passage of time, enabled by the invariant temporal structure of the mechanism is, in a sense, already taken care of by the updating of these estimates about worldly causes, becoming an inherent feature of the system at each level. When updates at all levels engaged in the generation of conscious percepts are considered together (more provocatively, when *we are the whole system*), we are left with a full-blooded phenomenology of temporal passage, including the notion of a specious present or window corresponding to what we consider to be "now" as well as past and future-directed information present throughout the system enabling a feeling of succession.

Ultimately however, it must be remembered the brain is not doing two things here. The brain is not making predictions about worldly causes that generate familiar percepts like trees and chairs AND making predictions about lengths of intervals, speed of passage, existence of passage, etc. This information is instead "built in" to the process of Bayesian updating itself, such that the system can't help but be temporally structured. This accords with the Kantian notion that experience cannot be had without *already being* temporal. While one can fail to perceive a percept or go blind and lose a source of sensory input, altering or perhaps ending predictions about the causes of that sensory input, the same does not hold with temporal experience. One cannot go "timeless" without ceasing to exist as a subject and agent (or so I have argued in Chap. 3). We should expect that time is a fundamental part of the mechanism responsible for any perception, and indeed any experience, whatsoever—a *sine qua non* of consciousness. As such time is therefore of an entirely different character and provenance from the everyday objects of consciousness.

Besides the retentional aspects outlined above, TIER is also extensional in that the constant Bayesian updates grounding minimal temporal experience occur at different clock times over the course of a measurable interval and that fact is relevant to the resultant experience. All the updates that concern current experience are integrated such that they form a finite but variable specious present. Within this window no particular processing level dominates in the sense of entirely constituting our perception, but rather they are all together implicated. The span of time through which the most relevant updates to perceptual expectations really take place determines the span of time we take to be now. There is therefore no crucial singular update for the perceiving subject, but instead a flexible window of time, consisting of all relevant updates about current stimuli. Within this window percepts are perceived as present, but the span of the experiential window is not itself the result of any explicit mental representation but rather a consequence of the actual operation of the mechanisms of consciousness in real time.

The "nesting doll" character of TIER mirrors temporal phenomenology as it is given. Henri Bergson, writing prior to Husserl, likewise emphasized a more complex temporal phenomenology than was commonly assumed. Bergson invites his reader to imagine a succession of musical notes. He then asks, "Could we not say that, if these notes succeed one another, we still perceive them as if they were inside one another [les unes dans les autres] and their ensemble were like a living being whose parts, though distinct, interpenetrate through the very effect of their solidarity?" (Bergson, 1889, p. 75 [101). He goes on: "One could thus conceive succession without distinction as a mutual penetration, a solidarity, an intimate organization of elements of which each would be representative of the whole, indistinguishable from it, and would not isolate itself from the whole except for abstract thought," (Bergson, 1889, p. 75 [101]). Thus no one note is "now" but depends on the others for its temporal character. The system, though essentially complex, has to be understood as a unity for its parts to make sense. Bergson's complexification of temporal experience as a heterogenous unity anticipates Husserl's tripartite scheme and later Merleau-Ponty's networks of retentions and protentions comprising the "field of presence," with "each present reassert[ing] the presence of the whole past which it supplants, and anticipat[ing] that of all that is to come…" (Merleau-Ponty, 1945, p. 488). Though the Phenomenologists would vehemently object to theorizing beyond what is given in experience, their general view of experience is perhaps best accommodated by a cognitive system that also exhibits complexity in the form of a recursive, multilevel hierarchy with feedforward and feedback connections between levels.

7.3 Minimal Temporal Phenomenology as Systemic Structural Feature

While the distinctiveness of temporal phenomenology is commonly recognized, it is equally common to stray from specific concern with minimal temporal phenomenology, understood as bare experiential succession and the temporal extension of the subjective present, and locate the relevant mechanisms at an implausibly high level of processing. This section will argue higher-order mechanisms are unrealistic as accounts of temporal phenomenology because such phenomenology is a ubiquitous feature in experience. On the flip side, TIER has the advantage of locating the most relevant features of the information processing for temporal phenomenology at the lowest functional level, i.e., in the way that expectations are formed and updated.

The distinctiveness of temporal experience vis-à-vis the other senses is well demonstrated by the absence of much sought-after internal clocks, as Anil Seth observes:

> For vision, we have photoreceptors in the retina; for hearing, there are 'hair cells' in the ear; but there is no dedicated sensory system for time. What's more, setting aside the circadian rhythm, which provides us with jet lag among other things, there is no evidence for any 'neuronal clock' inside the head which measures out our experiences in time—and which would in any case be a prime example of what Daniel Dennett called a double transduction, in which a property of the world is re-instantiated in the brain for the benefit of an assumed internal observer. (Seth, 2021, p. 134)

Because there is no evidence for a dedicated "sense of time" as there is for vision, and considering the *a priori* reasons for thinking there cannot be, we have plenty of motivation to look at the workings of the overall cognitive mechanism in order to account for temporal phenomenology. Seth goes on to assert, citing the work of Warrick Roseboom, that rather than "sense" time "we instead infer time based not on the ticking of an internal clock, but on the rate of change of perceptual contents in other modalities," (Seth, 2021, p. 134). Like Dayan et al. (1995), Roseboom et al.'s study involved an artificial neural network acting as a predictive engine called a Helmholtz machine: in this case a "feed-forward image classification network" designed to functionally simulate the human visual system (Roseboom et al., 2019). By analogy with this successful AI system, Roseboom et al. hypothesize that inferences play a role in human duration estimation.

However, it is not right to say, as Seth does, that "inferring time" explains minimal temporal phenomenology, because the latter should properly be seen as a precondition for higher-order temporal reasoning *about* that minimal phenomenology. It is only after there is perceptual change that a rate of change can be inferred from information about that rate. In other words, there must be experience over time—we must experience a duration—before we can estimate the duration of that experience. Thus, minimal temporal phenomenology is not the result of any specific inference(s), but rather a *consequence of the ongoing inferential processes* of the Helmholtz Machine. The structure of the mechanism rather than the informational content is paramount.

Hohwy, Paton, and Palmer (2016) endorse a similar but different higher-order thesis concerning experiential flow. They claim that "the sense of temporal flow in conscious perception stems from the probabilistic inference that the present cannot be trusted," (Hohwy et al., 2016, p. 315). I consider the thesis to be higher order because they claim that "for any level of perceptual inference, which proceeds according to its own learning rate, a higher level belief about precisions (i.e. a hyperprior) is needed to modulate that learning rate according to how volatile the context is expected to be (Mathys et al., 2012)," (Hohwy et al., 2016, p. 318). It is this (variable) hyperprior about precision of prediction errors that is meant to determine our perceived rate of temporal flow. On their view, it is ultimately expected precisions that determine the level of distrust there will be regarding the current sensory input. However, this higher order view of flow is indebted to the structural features and ongoing operation of the predictive mechanism for its raw materials:

> We argue that the sense of flow arises as the system gives up one hypothesis and settles on a new one due to its propensity for distrusting the present. The current prior decreases and this makes it seem like we are inexorably pushed forward. The window of the specious present moves forward because the system expects change and therefore down-regulates the current input. The sense of flow thereby occurs as a property of the internal workings of the hierarchical generative model. This goes beyond just predicting what will happen, which could happen in a non-hierarchical Bayesian perceiver with no sense of flow, and it does not rely on being entrained by the actual changes in the world's hidden causes. (Hohwy et al., 2016, p. 330)

Hohwy, Paton, and Palmer's proposal bears similarities to TIER, but also differs in significant ways, especially concerning theoretical commitments and explanatory focus. TIER is specifically concerned with a different explanatory target: minimal temporal phenomenology understood as the specious present and succession. As may now be clear, the suggestion is that "flow," understood as replacement, or, if you like, bare experiential succession, "arises as the system gives up one hypothesis and settles on a new one," (ibid.). Importantly however, the "sense of flow" mentioned by Hohwy, Paton, and Palmer, and "experienced flow" are not necessarily the same thing. This has been remarked upon on countless occasions in the literature on temporal phenomenology, most famously encapsulated by the Jamesian Doctrine, which Seth formulates in the following way: "*perception of change* is not the same as *change of perception*," (Seth, 2021, p. 133). It's important to note that what we get out of the model revision—the giving up of hypotheses for new ones—is *mere* experiential succession. Whether or not a subject is aware of this mere experiential succession as such may well be a question that requires appeal to metacognitive or higher-order hypotheses *about* what the system is doing. This question, which many authors like those discussed above seek to address, already goes beyond the basic question of how we come to have temporal experience in the first place.

Another important difference is that TIER makes no commitments to the content of high-level priors. *Why* one hypothesis is replaced by another is less important to this explanation than *that* it does. Hohwy, Paton, and Palmer may be right about the higher-level expectation a subject has to "distrust the present." However, we might worry this claim smuggles temporal language into a system that doesn't yet have a "present" until it is constructed through the operation of the predictive engine. While the claim might seem to invoke an objective metaphysical present, such a commitment seems unnecessary if we reframe the claim as the notion that the system has a higher-level expectation of volatile input. This alternative interpretation of Hohwy, Paton, and Palmer's hyperprior would still remain compatible with everything I suggest here.

Lastly, and most importantly, according to TIER the specious present does not "move forward" because the system expects change and therefore downregulates the current input, even if that is in fact something that happens. The present rolls "forward" just because the hypotheses are updated. The contents can even remain the same. Perceptions can last some amount of time for the subject even when not changing. Were it not for the updating of the hypotheses, on this view, there would be no specious present. Suppose the current input is not downregulated, but instead hypotheses are continually generated on the basis of that current input, assuming it is somehow retained by the system indefinitely. From the subject's perspective, there would be an unchanging experience of a certain duration. But in such a scenario, there is still temporal experience and a constructed present for a subject.

Of course, in answering why the hypotheses should change at all rather than leave us in a static non-experiential state where only one situation is ever inferred and thus only one situation exists for the subject, we could appeal to a higher-level expectation that the world changes and hypotheses must be updated. However, we could also dispense with the higher-level expectations of change and remain minimally committed to the idea that hypotheses simply change, perhaps because that's what being a biological organism in constant physical flux requires. Indeed, as Hohwy, Paton, and Palmer recognize, in the context of dynamical systems theory, transitions driven by dynamical instabilities can lead to a tendency to switch between states in any case (Hohwy et al., 2016, p. 323, n. 2).

The differences in emphasis between the current view and that of Hohwy, Paton, and Palmer can best be seen in the way the latter undertake to explain the phenomenon of binocular rivalry. Binocular rivalry is the phenomenon that occurs when different visual stimuli are presented individually to each eye (Wolfe, 1983). Rather than blend them together as one percept, the visual system appears to switch between the stimuli over time. One explanation for this phenomenon is that the combination of the stimuli as one percept is inferred to have much lower probability than their independent existence. Thus, only one percept at a time can be perceived. However, each hypothesis is equally good concerning the causes of the sensory stimuli, so there must be some further explanation about what pushes the system to switch between them. Hohwy, Paton, and Palmer point to the hyperprior we have about the untrustworthy present as the reason why we switch between percepts rather than settling on one (Hohwy et al., 2016, p. 325).

However, it seems that the more significant claim for the overall temporal character of consciousness, embedded as an assumption within the claim about the hyperprior, is that the revision of the expectation is the operative force behind the percept switching, whatever it may be that drives the expectation to be revised. This seems to be the heart of Hohwy, Paton, and Palmer's position, although it is underemphasized in comparison to their commitment to the importance of hyperpriors about precision. The central appeal to expectation revision should be how any predictive approach ultimately explains the phenomenon of binocular rivalry. As it is with binocular rivalry, so it goes with all successive experiences, even if the change of content is less immediately jarring.

Hohwy, Paton, and Palmer are quite right to point out that the "internal workings of the hierarchical generative model" are critical for understanding flow. I would argue these structural features of the mechanism are also critical for other aspects of temporal phenomenology, like the character and duration of the specious present. Recognizing this sets up an important contrast to traditional ways of thinking about these issues: temporal phenomenology is *not* mere representational content (or not *entirely*) but instead has to do with the way this content is structured by the operation of an ongoing cognitive mechanism.

7.4 Related Accounts

Rick Grush has proposed a related though, as we shall see, quite distinct model of the information processing mechanism responsible for temporal experience. He calls this the Trajectory Estimation of Model or TEM. This model can be seen as a way of grounding retentionalism in its computational implementation. Such a model is meant to show us how the basic phenomenological picture, following retentionalism, could arise from a predictive mechanism physically instantiated in the brain. The important thing that Grush demonstrates for our purposes is that the details of this implementation can inform our understanding of the phenomenology or higher-level model, while also explaining temporal experience in a more sophisticated way than the broad strokes of higher-level classic models[15] allow.

TEM is meant to model temporal windows of approximately 200 ms ("a lag and reach of on the order of 100 ms each, for a total temporal magnitude on the order of 200 ms,"; Grush, 2006, p. 444), which we can think of as comprising the specious present.[16] Grush identifies a window of this length for two reasons: it seems likely that such a timeframe would be that within which there is no "well-defined finish-line for information being propagated through the processing mechanisms of the brain" (Grush, 2016, p. 8), in keeping with Dennett and Kinsbourne's multiple drafts approach (Dennett & Kinsbourne, 1992), and because of data from studies on the timeframes of temporal illusions, especially the postdictive effects (Grush, 2005).

Grush attempts to ground phenomenological features of our experience in features of the brain's information processing system. In TEM, we are to think of perception at any given time as consisting of a trajectory estimate,[17] which is to say, not just the perception of an event but a perception of where that event came from and where it is going. More specifically, Grush appeals to what he calls "smoothed estimates" which correspond to future events and "predicted estimates" which correspond to past events, and these are meant to be the information processing underpinnings of protentions and retentions. Furthermore, so-called "filtered estimates" combine prior knowledge with incoming data to arrive at an estimate of what is going on. Perception as a whole is supposed to be "embodied in the state of the model, which is the perceptual representation of the environment" (Grush, 2008, p. 152), including all of the kinds of estimates listed above.

The nature of these estimates can be fleshed out in greater detail. In the case of filtered estimates, "the perceptual system's knowledge about the represented domain, in particular how the domain typically behaves and other statistical regularities" combines with "sensory information about the domain picked up by the sense organs," (Grush, 2008, p. 152). Both factors, i.e., prior knowledge and incoming sensory information, are capable of overriding the other depending on circumstances. The filtered estimates form the core "data" of the percept, specifying what is represented as likely to be happening now and act as the basis for projections into the future and past. Filtered estimates differ from what Husserl called primal impressions in that they are not raw incoming sensory information, but rather employ a complex of prior knowledge in combination with incoming sensory information to make an estimate. Because of this, information from the past is already implicated in the construction of such estimates, although the estimates are directed at the current environmental (or internal) situation.

---

[15] The classic models discussed in Chap. 4.

[16] Within such windows Grush thinks that events exhibit a "B-ish" temporal character, which is to say, they are not experienced as "now," "past," or "future," but only in relation to each other as "earlier than," "later than," and "simultaneous with," (Grush, 2016).

[17] This trajectory estimate is described by an ordered tuple of events, which is to say, a sequence where the order of the elements matters, so that succession is represented (Grush, 2006).

For Grush, the technical term "predicted estimate" refers to a specific, iterative process. Predicted estimates are not about what is currently going on but about what will happen later. The system takes "knowledge about how the domain typically behaves" and combines this with "a representation of the current state," which is to say the original filtered estimate, and then "project[s] ahead in time," (ibid.). Insofar as predicted estimates involve predictions about the future, they can be seen as computational-level analogues to Husserl's protentions with their future-directed intentionality.

On the other hand, smoothed estimates, rather than being directed towards the future as is the case with predicted estimates, are directed towards the past. Here the filtered estimate ("representation of the current state") and knowledge "about how the domain behaves over time" combine to form an estimate of what was happening at some previous time (Grush, 2008, p. 153). Grush claims that this "smoothing" process can actually produce better estimates than the original filtered estimates by appealing to "back-tracked versions of later estimates," (ibid.). In other words, this process can iterate backwards in a similar way to how predicted estimates iterate forwards, projecting the state of the system into the past based on its current state and making estimates based on that backward-projected state. Smoothed estimates, insofar as they are past directed, can be seen as the computational-level analogues of retentions.

The upshot of these three processes is the heart of Grush's Trajectory Estimation Model: "By combining filtering, smoothing and prediction, these mechanisms are able to produce, at any time *t*, an estimate of the behavior of the represented domain over any interval, [although] as the times involved in the predicted or smoothed estimates get farther from the present time, their accuracy becomes less and less reliable, so any useful interval will often be a brief one centered on the present time," (ibid.). Presumably, beyond this window of reliability the experience would no longer be considered a part of the specious present, which Grush has indicated is probably something on the order of 100 ms in either direction, based upon postdictive effect studies. Notice, however, notwithstanding the increasing levels of inaccuracy, there is still the uncomfortable sense that, as with retentional models generally, there is nothing in principle preventing the system from extending indefinitely in both past and future directions (see Chap. 6).

Supposing there is a boundary surrounding what we take to be present, what happens beyond that? When TEM is run online, i.e., concurrent with and directed at the temporally immediate environment, the model is supposed to result in the phenomenology of perception as we know it. However, Grush also maintains the model can be run *offline*, i.e. not concurrently, "in order to produce expectations of what the modeled domain might do in this or that circumstance (or if this or that action were taken by the agent)," (Grush, 2006, pp. 441–442). Thus, Grush is claiming that this model is also at work during imagining and remembering, but in a slightly different way, with different kinds of representations (what he calls "conceptual representations," (Grush, 2006, p. 447). These conceptual representations get their name from "involving concepts of processes that span potentially large intervals [beyond 200 ms]," (ibid.). Conceptual representations are contrasted with "perceptual representations" that are confined to the space of the specious present, i.e., the approximately 200 ms temporal interval that constitutes our "now." According to Grush, this difference is also supposed to explain the stark phenomenological difference between perceptions of, e.g., music playing now and music we expect to hear in five seconds or that we did hear five seconds ago (ibid.).

The estimates at work in TEM can be seen as representations, and the particular characteristics of each accord with the features of a more or less retentional model of the phenomenology of temporal experience. TEM has affinities with Husserl's tripartite structure of the specious present but is not completely analogous. One important difference is that, in TEM, "perceptions don't 'sink back' into retention, as Husserl put it. Rather, "the entire interval estimate is constructed anew each time," (Wiese, 2017, p. 6). Another difference is that, whereas Husserl seemed to think his analysis applied to larger timescales, Grush restricts his to a 200 ms interval. As previously mentioned, Grush invokes two different representations for different timescales: "perceptual representations" for sub-200 ms and "conceptual representations" for larger intervals (Grush, 2006, p. 447). Furthermore, Grush objects to the way Husserl "privileges a now-point and gives it the name *primal impression*. The trajectory estimation model does not privilege any of the phases within the temporal interval," (ibid., p. 448). Grush worries admitting primal impressions means, "temporal illusions should not be possible," because the primal impression is invariant (ibid.). Due to filtering in TEM, "as time progresses, the *entire trajectory* is re-estimated, with the consequence that some parts of the estimate can be changed," and this he thinks can solve the problem of illusions (ibid.).

Grush specifically favors a representation-heavy retentional model in order to account for temporal illusions (Grush, 2005, p. S210). As we have seen, some of the most paradoxical temporal illusions are the postdictive effects. These illusions seem to require that time is an object of representations, because we seem to *mis*represent time when they occur. As Grush puts it, "something has to be *represented* for it be *mis*represented," (Grush, 2005, p. S217). For representational views of time, an illusion is just a representing of time *as* being a different way than it "actually" is. In TEM-style models, an illusion is just a statistical irregularity in the signals coming from the environment that does not match up with expectations, resulting in inaccurate event estimations (ibid.). These inaccurate estimations alter our experience of *now* because they are constitutive of it. Grush maintains that the model retains "openness to revision for, say, 100 ms," (ibid., p. S216), which is why postdictive effects are not observed outside this timeframe (Fig. 7.3).

7.4.1 The Hierarchical Extension to the Trajectory Estimation Model (HiTEM)

TEM is supposed to ground a phenomenology that accords with something like Husserl's retentionalism, albeit with modifications (for example, the removal of primal impressions). Wanja Wiese's Hierarchical Extension of the Trajectory Estimation Model, or HiTEM, is meant to be a modification of TEM that

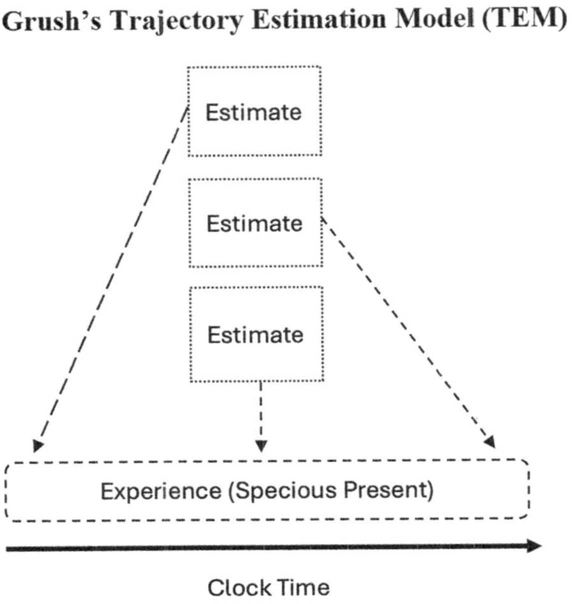

Fig. 7.3 Illustration of Grush's trajectory estimation model, inspired by Grush (2005) and Wiese (2017). Simultaneous estimates (represented by arrows) are made of the probability of events at different time-phases. These estimates are not themselves extended in time, or at least not in a way that is explanatorily significant for any reason. The lack of significant temporal extension is represented by the dotted borders around the estimates. Together these estimates constitute the specious present. Experience is presented here with a dotted border to represent that, in line with retentionalism, it is experience *as of* a duration but is not necessarily itself extended in clock time (see Wiese, 2017, p. 5 for another diagram of TEM)

accommodates Grush's model within the predictive processing framework. Both of these views are meant to do justice to retentionalism and differ from TIER in their denial of the explanatory relevance of any real temporal extension and their emphasis on the representation of intervals. However, unlike TEM, Wiese's HiTEM shares with TIER an appeal to a hierarchical predictive system.

Wiese found Grush's trajectory estimation model unable to answer what he calls the "interface question," and sees his "hierarchical extension" as the remedy. TEM's large interval conceptual representations are structured according to the tensed A-series (i.e., they are past, present, or future), while perceptual representations are structured according to the tenseless B-series (earlier than, simultaneous with, later than). We are thus left with two different kinds of representations at two different timescales. From a phenomenological standpoint, however, we seem to have a continuous and enduring experience of time with no discrete boundaries or distinctions between timescales. The interface question is therefore: "How are perceptual representations of sequences integrated with conceptual representations of sequences" or, more simply, "How can I experience recent events as being seamlessly connected to present events?" (Wiese, 2017, p. 7).

Building on Grush and TEM to answer the interface question, Wiese appeals to hierarchical predictive processing. Rather than two kinds of representations for different timescales, Wiese's hierarchical extension of the trajectory estimation model takes it that "there is a hierarchy of temporal wholes" with each responsible for a different timescale (ibid.). Rather than a difference in kind between conceptual (long-interval) and perceptual (short-interval) temporal representations, in Wiese's model these timescales are just represented at different levels of the hierarchy, because each level is responsible for a different timescale.

Wiese also introduces "mediating" representations, which are responsible for "represent[ing] features as being instantiated during an interval which is longer than that identified by Grush [200 ms]: it represents features as having been present in the recent past, as being present now, and as continuing into the near future," (ibid., p. 15). Our ability to perceive timbre and rhythm is supposedly dependent on such mediating representations, which allow us to experience an event as having been present before just now and expected to continue afterwards.

Our holistic experience, according to HiTEM, is composed of all of the predictions of events made at each level in the hierarchy. If the subject detects "an increase in prediction error, an event boundary is inferred, and the event model is updated," (ibid., p. 19). This way of inferring event boundaries is essentially identical to that proposed by Zacks' Event Segmentation Theory (EST), although EST operates on a timescale of seconds, rather than a fraction of a second (Tversky & Zacks 2013; Zacks et al., 2007; Reynolds et al., 2007).

Beyond predicting singular events, the hierarchy also allows that each level is predictive of the levels above and below, such that, e.g. a higher-level representation of rhythm can predict "the *timing* of individual notes" and "emotional responses can be predictive of the *key* in which a melody is played, and the key can be predictive of intervals in the melody," (Wiese, 2017, p. 19). Reciprocal predictive power between levels is a feature of predictive processing more generally, which Wiese here exploits to explain our perception of the interconnectedness of temporal scales. This attractive feature and Zack's EST concerning event boundaries are also both compatible with TIER (Fig. 7.4).

7.4.2 *Problems with TEM/HiTEM*

TEM and HiTEM share significant conceptual shortcomings. The first of these is a failure to answer the interface question, at least in its jargon-free formulation: "How can I experience recent events as being seamlessly connected to present events?" (Wiese, 2017, p. 12). HiTEM's mediating representations shift the problem one step further away but the problem remains. Whereas Grush had the problem of integrating conceptual and perceptual representations, Wiese now has the problem of integrating conceptual, perceptual, and mediating representations. What mechanism is

Wiese's Hierarchical Extension of Grush's Trajectory Estimation Model (HiTEM)

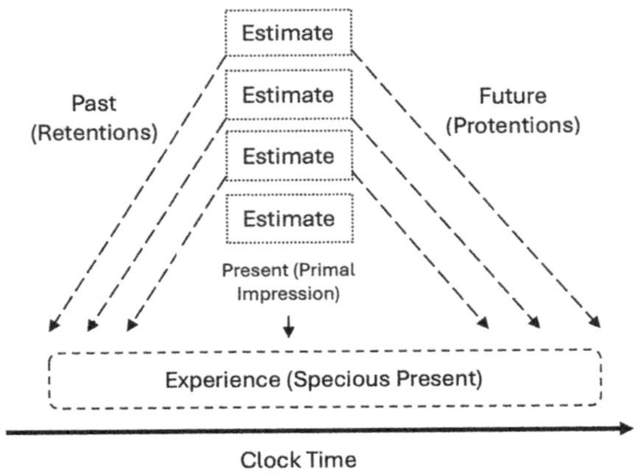

Fig. 7.4 Wiese's hierarchical extension of Grush's trajectory estimation model, with representations for various timescales at each level in a reciprocally predictive hierarchy. As with TEM, the arrows represent estimations (diagram inspired by Wiese, 2017; see Wiese 2017, p. 12 for original)

there for seamlessly connecting mediating representations at different timescales to each other or to perceptual and conceptual representations? Adding more representations appears to be a "tortoises all the way down" kind of solution.

A bigger problem with TEM and HiTEM is that these models see time as the object of representations. In this context, and as Grush has defined it, "object" means "*something conceived by* or *presented to/in* a conscious mind" and therefore need not be concrete or real (Grush, 2006, p. 419). Grush and Wiese are committed to a view of time as an object that is represented more or less like any other: TEM (and HiTEM) "explicitly represents the temporal aspects of the modeled domain just as typical schemes represent the nontemporal aspects," (Grush, 2005, p. S210). This representationalism about time is in line with other internal modeling approaches, wherein, "the system has an internal model of the perceived entity (typically the environment and entities in it, but perhaps also the body), and at each time *t*, the state of the internal model embodies an estimate of the state of the perceived domain," (Grush, 2006, pp. 441–442).

The problem is that the very possibility of representing an object is predicated on the preexisting temporal structure of experience. Objects may be represented as having certain qualities, for example, they might be represented *as* changing. However, insofar as temporality is a prerequisite for experience, it should be the case that on a more fundamental level such representations are temporally structured. This is why the model should insist that representations inherit the temporal structure of the physical system they are instantiated by.

7.5 Advantages of TIER

Rick Grush objects to the widespread tendency to take time as, "its own best model, indeed its only model," (Grush, 2005, p. S210). By this he means that, "the representation of succession is accomplished by a succession of representations," (ibid.). However, there is an important intuition motivating such claims that Grush overlooks, which makes them even more compelling than the roboticist Rodney Brooks' more famous general claim that the world is its own best model (Brooks, 1991, pp. 139–159). Besides the idea that it seems evolutionarily superfluous, i.e., inefficient and therefore potentially maladaptive, and theoretically unparsimonious[18] to construct a representation of something that's already in the world to be exploited, there is also the deep Kantian insight that time, like space, is very different from other things. In fact, time is not a thing at all. Viewing time as an object of representation trivializes the role that time plays as the necessary structure or field of all other subjective experiences.

TIER accommodates a phenomenology that is temporal at all grains. There is no scale or kind of experience that lacks temporal structure, nor could there be. Strongly representational models are in the uncomfortable position of claiming that time and other objects of perception are represented in the same way and then somehow combined such that the latter are subsumed by the former. If the levels of the hierarchy in predictive processing are the building blocks of consciousness, we should expect each level to exhibit the temporal structure that conscious experiences depend upon—a temporal structure that is already present at the subpersonal level, under the threshold of awareness, and is carried through to the personal.[19] Conveniently, predictive processing already comes with such features.

Kant thought time must be a precondition of any experience whatsoever. Time is not derived from experience of the world. In the current proposal, Kant's position is taken to heart. Time, or the flow of time and the experience of the present moment, cannot be the result of the same process that gives rise to particular appearances, i.e., objects of experience. Whereas in the latter case predictive processing appeals to predictions of hidden causes in the world, we cannot admit that time results from predictions of causes of input impinging on a "time sense apparatus," or something similar, for this would allow that experiences can be had apart from time, which is

[18] At least, contrary to Ockham's Razor as Russell formulated it: "Whenever possible, substitute constructions out of known entities for inferences to unknown entities," (Russell, 1924, p. 160).

[19] Predictive processing is often taken to be a general organizing principle for all nervous systems and is sometimes even extended in ways that might be relevant to all forms of biological life, especially through appeal to the so-called "free energy principle" (Parr et al., 2022). If anything like this is right, then even non-sentient creatures with a similar cognitive setup would still have a temporally structured non-conscious cognitive system waiting to emerge were they complex enough to exhibit consciousness as we understand it. Alternatively, some might see this as a consideration in favor of expanding the domain of sentience to all those creatures that are organized according to such principles. I will remain agnostic about these contentious issues here.

impossible. Instead, TIER sees time as an inherent feature of the processes responsible for perception, and it is inseparable from those processes. The flow of time falls out of the Bayesian inference mechanism in virtue of its constant updating of predictions, but time is not itself an object of those predictions.[20]

TIER also avoids the so-called "interface question" by eschewing representations altogether. There is no question of how representations are seamlessly integrated when representations are not the vehicles of the relevant information for minimal time-consciousness, considered here as experiential succession and the experiential interval of the present. Instead, the predictions and error signals provide temporal cues at each and every level which are incorporated into updates that integrate seamlessly because that is all there is to provide the relevant information about temporality. This is analogous to perceiving a seamless integration of information despite blinking. No "extra" representations are required to fill in the space—the extant input does so automatically because the absence of information is simply not experienced. As Cyriel Pennartz explains in the case of blinking, "the lack of sensory input during an eye blink need not be actively represented in the visual system. In fact, the lack of such a representation may be the very cause of not noticing this gap, thus accounting for a feeling (or illusion) of continuity," (Pennartz, 2015, p. 51). Similarly, no extra temporal information or representations are necessary to integrate the information, because "...in the short run, spatial or temporal continuity in our experience may come about because of a lack of information on discontinuity," (ibid.).

On timescales much beyond the specious present, we might observe that, in fact, the *explanandum* of the interface problem is much less compelling than Wiese had indicated, for there actually *isn't* seamless phenomenological integration between sub-200 ms timescales and beyond. Times on either side of the subjective "now" do not feel as if they are now and do indeed feel quite disconnected. The phenomenology of memory and anticipation are distinctly different than the phenomenology of perceiving the present. However, within the context of the specious present, the suggestion that different levels may handle different timescales is spot on. Contrary to Wiese's view, however, TIER maintains this occurs because different levels update over time, in so doing targeting objects over different timescales.

[20] I leave open the possibility that we may represent time in a derivative way. It is after all a fact that we are capable of representing time by counting, using watches, etc. We can judge lengths of time, break it into units, and manipulate our representations of time to make calculations. However, I see all of this as a higher-order phenomenon involving some degree of abstract thought which is not present in our most basic, pre-reflective perception of the world.

7.6 Internal Clock Models

So far, I have said fairly little concerning internal clock[21] models as a potential basis for temporal phenomenology. Such models are historically popular in the field of psychology and have also been analyzed and defended philosophically, notably by Carlos Montemayor (2013).[22] I do not dispute that something that can be called a "biological clock" exists in human beings and many other living beings that regulates bodily rhythms. In mammals we find this system centered on the suprachiasmatic nucleus of the hypothalamus and there are pacemakers operating at the cellular level as well. I would also not dispute that human beings are endowed with the ability to represent and judge periods of time, and it may even be the case that there are distinctive "clock-like" cognitive mechanisms that facilitate this ability. Indeed, the hunt for these mechanisms and their neural substrates is ongoing.

However, I do not think that the clock metaphor offers much explanatory purchase on our basic temporal phenomenology—that is, the experiential present. This is because, as mentioned earlier, despite researchers' best efforts to determine the physical location of the relevant clock(s), whether that be isolated to a specific area of the brain or distributed throughout, no obvious contender for an internal clock has yet been identified. Were something like this found to be localized in a particular part of the brain we would also expect that cases of brain damage would arise in which people completely lose their temporal experience, but this has not been observed.[23] More importantly, the very concept of the internal clock as an explanation is inappropriate here, because in order for a clock to be meaningful as a marker of duration there must already be a subject to interpret the data of the clock. Without such special interpretation by a subject there is nothing particularly temporally interesting about a clock above and beyond any other process that takes place over

[21] The discussion here will remain high-level. Clocks can be periodic, i.e., the way we measure hours or days in repetition, or accumulators, like stopwatches. Various models have been proposed and Montemayor (2013) is an excellent philosophical resource on this subject, but my concern in this section is with the general explanatory strategy of appealing to internal clocks to account for our experience.

[22] Montemayor argues more specifically that, "the metric outputs of the [internal] clocks determine representation spaces of possibilities that structure and organize the behavior of an organism, by guiding it successfully in the environment and by helping it decide on the best course of action. These representations form reliably produced (perceptually driven) beliefs about duration." (Montemayor, 2013, p. 141).

[23] What *is* observed, however, is that dysfunction and damage to specific parts of the brain such as the striatum can affect a subject's ability to measure or "tell" time, and this is a common problem in neurodegenerative illness (Malapani et al., 2002). However, this is not the same thing as losing experiential time but instead a failure to properly store memories, recall memories, or make judgments.

time—grass growing, paint drying, radioactive decay, etc.[24] So, the reasons for skepticism are both empirical and conceptual.

There are further arguments, however, for why clocks are probably not helpful that I have not yet mentioned. Call the first the "argument from preexisting clocks". Given that we *actually have clocks available to us already* in the form of the biological clocks in both the brain and the cells of our body, the fact that these are *not* directly influencing our ongoing perception seems suggestive of the fact that such mechanisms are somehow ill-suited for this role. In fact, the central neural pacemaker in the hypothalamus can be damaged or removed without deeply affecting the temporal phenomenology of the present. If this sort of mechanism were the right kind to ground experience, and given that it is so evolutionarily primitive, why would the brain not utilize these preexisting clocks? Of course, this argument is not a proof, and one might want to respond that evolution is unfathomable to mortal minds. But it remains the case that the sought-after sense-organ, the cousin of the biological clock that is meant to ground temporal phenomenology directly, has not been found. Therefore, on balance, the considerations laid out here are at least suggestive that perhaps the reason it has not yet been identified is because it never will be.

A second argument comes from the reliance of clocks on metrical representations of duration, with the content of these representations generally thought to have veridicality conditions (Montemayor, 2013, p. xii). Call this second argument the "argument from the absence of veridicality conditions." We tend to think an accurate clock gets it *right*, while a broken clock gets it *wrong*. What does it get right and wrong? The actual amount of time that has elapsed in the world in one's particular reference frame. However, when we think about how we ordinarily experience the world, there is no such measure is to be found. Time may feel as though it is passing, but until we begin to measure it, we are not representing durations in such a way that those representations could even be accurate or inaccurate. We rely on essentially arbitrary metrical representations, seconds perhaps, to count out the time, and then the final representation of duration that our judgment determines would seem to have veridicality conditions. If we check our count against the stopwatch, we might say our representation is roughly accurate or false. But if we do not do this, but merely experience the world normally, it is hard to see in what sense we are measuring anything. There are no units of measurement and no particular basis for comparison. As a result, we lack obvious veridicality conditions. Feeling like it was "a long time" is already the beginning of measurement, but such an evaluation is generally absent unless we are prompted in some way. Such a prompt switches the subject from a mode of mere experience to judgment. It is via judgment, through

[24] Indeed, decaying isotopes *are* used as clocks, for instance in radiocarbon dating, but this does not imply that their "clock" status predates their use as such. Radiocarbon dating also requires complex interpretation and information about the state of the Earth and its atmosphere over time. Appealing to the representations of internal clocks is not made more sensible than appeal to any other dynamic process just because the interpretative process associated with them is simpler and more familiar to us owing to the ubiquity of man-made clocks.

comparison, that we can measure, and it is here that I think it is plausible to say we represent time. These representations may have veridicality conditions. But it is only when judgment is involved that it works that way. Thus, the absence of veridicality conditions in the normal non-judging experience of time would seem to point away from representations of time being necessary for the bare experience of the present, let alone being explanatory of it.

There is probably a strong evolutionary incentive to make accurate judgments of duration. However, there is no such evolutionary incentive for all of experience to be based on the kinds of representation that are employed in such judgments. To take an analogy, while there is an evolutionary incentive to accurately judge distances in space, it would be wrong to conclude on that basis that internal rulers or tape measurers in the brain form the basis for our visual phenomenology in its entirety. Furthermore, as Kathleen Akins (1996) has pointed out, sensory systems are *narcissistic* in the sense that they are concerned primarily with the survival of the animal and *not* with accurately representing information or measuring the world. Now of course I am actually arguing that time is not a sensory system at all, but if a proponent of internal clocks thinks of these as the temporal analogues of other sensory organs, the narcissism of our existing sensory systems would be something to grapple with, as veridicality conditions in that case, if there were any, might only arise incidentally. None of this is to say, however, that we don't routinely make judgments that do have veridicality conditions. It is only to say that these judgments are just that: judgments. We should not think that the representations involved in our judgments offer us a key to a deeper explanation of the experience the judgments are made over.

Now one might argue that perhaps at a subpersonal level there are comparator mechanisms like clocks ticking away and measuring against objective time without our awareness, and that it is these mechanisms that inform perception. But again, such mechanisms are not forthcoming and those that we *have* found do not exert the kind of direct influence over the phenomenal character of immediate experience that we might expect. Could some hypothetical mechanism of this kind regulate experiential rhythms in the same way as it regulates bodily rhythms? Perhaps it is possible. However, at this point experience would *already be there* to be regulated, and as I have argued, that experience must already be temporally structured. Furthermore, if something like TIER is on the right track, we can tell a fairly decent story about how experience comes to have such a temporal structure without explicitly resorting to some internal clock-like mechanism apart from the basic cognitive machinery responsible for consciousness itself.

Granted, insofar as the workings of the circadian clock affect the body, they affect the experience of the animal indirectly via hormonal mediation, and so in that sense the rhythms of the biological pacemakers can be said to have some relevance to our felt experience, for example, when and to what extent we feel tired or hungry etc.. Additionally, we cannot rule out other clock-like mechanisms in cognition. We are able to judge durations. In order to do this, we can do things like count. In a sense then, we are able to avail ourselves to a kind of mental clock and make judgments based on this clock. This is important for timing actions and surely adaptive.

However, these sorts of clocks, whatever the answer to the independently interesting question of their neurological and cognitive mechanisms, will not be the basis for the subjective experience of the present, which is to say, time-consciousness.

The deepest conceptual problem with explanatory appeals to internal clocks is that the functioning of internal clocks *qua* clocks presupposes the subject they are enlisted to account for, insofar as the subject must *already* have temporal experience in order for the information carried by the clock to be interpretable. Earlier I offered a similar line of reasoning against the idea of "temporal representations," which take time to be a kind of content in the same way that other perceptual representations allegedly have other forms of sensory information as content. But though we *can* represent time, it is a mistake to think that, through such temporal representations, we come to perceive time itself. Once again, it is more sensible to think we do not perceive time at all, but instead perceive temporally. I have suggested that we do this by running a predictive engine that continually constructs a present experience and that the predicted causes of stimuli that are experienced are structured in time due to the operation of the predictive mechanism. This is just what it means to experience temporally according to TIER.

Nonetheless, it is still true that temporal representations are everywhere in our lives. We represent times in many ways every day, and it is quite useful to us that we can do so.[25] Some temporal representations may modify aspects of experience, as Husserl suggested, such that, for instance, protentions are felt as future and retentions felt as past. These may be understood as representing a particular time-phase, with the sum of these representations constituting the whole event that we experience in the present. However, even here it is events that are represented in a particular temporal way, but not time that is represented as the content of the representations. In any case, I think this representational understanding should not be taken overly literally in the context of TIER. Because of the way the underlying cognitive mechanism works, I think instead the continuously changing relation of the subject to the objects of experience should be seen as the most basic process in the construction of the present. Representations are thus not similarly suitable as a fundamental explanation, or so I am suggesting.

Of course, not everyone will resonate with this clock-antipathy. The internal clock model has remained a live and sometimes even assumed theoretical posit in psychology for many years, even if it is perhaps less popular now than it has historically been. On the fundamental question of whether we perceive time or whether we perceive temporally, TIER certainly diverges from an internal clock model. However, it is also the case that the explanatory targets of internal clock models often differ from my concerns here, and for some of these targets, e.g., explaining

[25] The phrase "5 seconds" is an easy example of such a representation, as is the sand in an hourglass, which is a temporal representation as long as someone understands it to be so, otherwise it is simply sand. The same is true for the plastic hands of a clock or the lights on an LED, though familiarity has trained us to think of their connection to time as somehow more deep and inherent than the sand on the beach, which yet moves through time in exactly the same way as any measuring device.

timing judgments in the context of timing studies, there is not necessarily an obvious incompatibility between the two views. I think it is probably because of how suitable the metaphor seems to be in accounting for bread-and-butter psychological experiments in this area (like timing tasks) that the clock theory has remained somewhat entrenched in psychology in particular.

I speculate that where serious proponents of internal clock models go wrong is in taking the theoretical posits of experimental psychologists too seriously, when these assumptions are better seen as ad-hoc stand-ins for the *unknown* mechanisms responsible for the behavioral data that they collect. Assuming such a mechanism in a more serious way than as a convenient stand-in is problematic firstly because the psychologist is assuming a mechanism that neuroscience has so far failed to substantiate and secondly because the nature of the experimental data makes the clock model appear to be a more plausible assumption than it really is. Psychological timing studies, though ostensibly straightforward and objective, frequently conflate three distinct things: (1) the subjective *report* of the temporal experience during a task, e.g., how long a subject says an event took, which crucially involves the subject's memory; (2) the subjective judgment made *about* the temporal experience, which is often assumed to be tracked by the report; and (3) the subject's actual temporal experience, which is perhaps thought to be accessible through retrospective subjective report or behaviors. However, it is not at all clear that the three of these need to correspond closely, and, in some cases, it is clear that they do not. We need only think of how routine, boring events may lead to subjective time dilation ("this commute is taking forever") in the moment but in retrospect may seem to barely register as happening at all.

Regardless of the conflicting empirical evidence, from a conceptual standpoint the above phenomena are simply not equivalent. It is therefore possible, and indeed, very likely, that independent or at least semi-independent mechanisms underpin each of them. The metaphor of the clock does a decent job accounting in particular for *judgments of duration*. It just so happens that judgments of duration are perhaps the most tractable experimental paradigm for psychologists to study, and thus might easily be conflated with temporal phenomenology writ large. I believe this is one of the main reasons for the appeal and staying power of the internal clock model, but I also believe this conflation, tempting as it is, must be resisted. TIER is not concerned with how we come to form judgments of duration, regardless of whether those judgments occur at an explicit level of awareness or not. Instead, TIER is meant to provide an account of temporal phenomenology in its minimal form, that is, our basic subjective experience of the present, which I have been calling time-consciousness. Ultimately, the appeal to clocks, either literal or metaphorical, cannot account for this basic subjective experience without assuming the existence of the thing it is trying to explain.

7.7 The Origins of Time-Consciousness

This book has until now mostly concerned itself with the way time-consciousness is for us as we find it in our current experience. In other words, the discussion has taken place on a *synchronic* level in that I am considering our present experience as it is right now. This section will begin to consider time-consciousness *diachronically*, that is, as it has developed over the course of deep history, in order to demonstrate that there is a plausible story about the emergence of TIER as an evolutionarily adaptive information-processing strategy.

It is sometimes said that "Time Flies Like an Arrow; Fruit Flies Like a Banana."[26] Not only is this a good joke, but it also helpfully connects the apparent arrow of time, understood here primarily as a thermodynamic gradient (Callender, 2017, pp. 132–133), and the evolution of life, in particular the self-sustaining impulse that leads flies to bananas. I will argue this connection is important to cognition and consciousness, particularly time-consciousness, because the pressures of our local environmental conditions, which include an apparent arrow of time, favor systems that are able to exploit these conditions appropriately in order to persist. The best way they can do this is through ongoing prediction enabled by local synchronic and diachronic variations in entropic conditions. These predictions themselves exhibit inherent temporal structure and must be revised repeatedly in complex environments.

The early universe was highly ordered in the sense of being extremely dense. It is thought that everything began in a state of extremely low entropy, where entropy is understood as the measure of disorder in a physical system. According to the second law of thermodynamics, systems generally increase in entropy (disorder) until reaching an equilibrium. Though rare, pockets of relatively lower entropy are an inevitable consequence of our position somewhere along a larger trajectory from low to high entropy. There are thus still remnants everywhere of our low entropy (highly ordered) origins available for us to exploit. The existence of these remnants is a critical enabler of our continued existence, whether on the scale of the galaxy, the solar system, or the planet, which contains abundant heavier elements and complex organic materials that can be used to build living systems. At some point, perhaps, all these physical systems, at every scale, will devolve into disorder, but for the moment they provide the ingredients for the temporary sustainment of relatively well-ordered systems like ourselves. Indeed, one way of *defining* what it means to be alive is the active maintenance of an improbably low entropy state against a higher entropy background that tends towards disorder (Seth, 2021, p. 198; Corcoran et al., 2020, p. 3; Parr et al., 2022).

---

[26] This quote came to my attention as the epigraph of the preface of Huw Price's *Time's Arrow & Archimedes' Point* (Price, 1996, vii), attributed to Groucho Marx. According to the *Yale Book of Quotations*, "there is no reason to believe Groucho actually said this," (Shapiro & Henderson, 2006, p. 498). The quote may have originated with Anthony Oettiger, who discussed its grammatical ambiguity extensively in a *Scientific American* article about computational syntactic analysis, in which he argues "no computer known can deal effectively with semantic problems of this kind," (Oettinger, 1966, p. 169). That piece contains other entertaining tidbits such as this: "It is after all, only an accident of nature, or for that matter merely of nomenclature, that there is no species of flies called 'time flies'," (Oettinger, 1966, p. 168).

The notion of a cosmic origin in a singularity and the idea of a universal thermodynamic gradient are fairly standard assumptions. The next step in the story is the recognition that the thermodynamic gradient is responsible for conditions that guarantee an informational environment exists that allows for adaptive exploitation of the pockets of relative order left over from the early universe's low entropy origins (Ismael, 2021, p. 76). These leftovers can be understood as exploitable informational *records*. Wherever there are records there is the possibility to use them to make inferences, either predicting or retrodicting.

A universe in equilibrium has no records—no way to determine its evolution over the dimension of time. A record could be a clump of matter hurtling through space, whose own origin can be inferred by its present velocity. Or it could be a compound emitted from an undersea vent or secreted by an organism. The entities that exploit these records are *themselves* records on multiple levels, most obviously as encoded by genetic material, but also in the way their bodies and behavior conform to the ecological niche they have developed in.

The example of the compounds emitted from undersea vents highlights two points. The first is that such compounds can literally be *food* in the sense that these bits of relative low entropy against their relatively higher entropy background help to sustain an even lower entropy pocket which consumes them (e.g., chemosynthetic bacteria). The second point is that it would be very beneficial for an entity to treat certain chemical substances as informationally loaded signals to predict the location of the source. In so doing, the entity can perhaps obtain more of the resource. Here is one answer to why fruit flies like bananas—the thermodynamic arrow of time has made it so. Fruit flies are drawn to bananas because they provide a chemical signal (e.g., fermenting sugar) that indicates the presence of life-sustaining compounds that are expected to aid homeostasis, which can be understood as the resistance to dissipation into the environment.

Not long after the emergence of life, there would have been an explosion of organisms in competition, each benefiting from literally consuming the others for the maintenance of themselves.[27] In this environment potential "records" multiply. It becomes important to sense and operate on records over multiple timescales, interpret them appropriately, and generate longer-term expectations. Though fallible, extrapolation to the future from records works better than any alternative to ensure continued existence, which is all that is required for its widespread adoption as an informational strategy. As Craig Callender observes, "[what we commonly call] time is that direction on the manifold of events in which we can tell the strongest or most informative stories," (Callender, 2017, p. 142). It is this informational direction, itself determined by the thermodynamic gradient, that then goes on to determine our subjective arrow of time.

[27] Peter Godfrey-Smith (2020), in his exploration of the origins of cognition, assigns great weight to the evolutionary arms race that occurred during the Cambrian explosion, a period starting roughly 540 million years ago when there was more oxygen available. The fossil record from this time indicates a proliferation of diverse bodily formats resulting from the transition to widespread predation from simple scavenging. These new pressures encouraged rapid evolution and ever more sophisticated possibilities for perception and action, which he sees as the beginning of something resembling minimal consciousness in animals like crustaceans with spatialized sensory systems, high mobility, and a need for self-recognition (Godfrey-Smith, 2020, pp. 86–88).

So far in this section we have been charting a broadly "biogenic" story of the origin of cognition (Lyon, 2006). As Matt Sims puts it, biogenic approaches think of "cognition as a process that underwrites an organism's ability to remain in thermodynamically improbable non-equilibrium steady states despite the tendency for all systems to move towards thermodynamic equilibrium" (Sims, 2021, p. 5). These approaches place a strong emphasis on how homeostasis is achieved by complex self-organizing systems that can actively change certain parameters to resist destruction by a hostile environment and see cognition as continuous with such activities (ibid.).

From the biogenic perspective, the facts of cognition and consciousness today are inextricable from their biological and ultimately physical origins in a world with very specific conditions. For organisms that inhabit such a world, their activities are performed embedded in positions along the thermodynamic gradient that are not time-invariant, as in idealized cases of physical systems, but exhibit a strong organism-relevant time asymmetry. It is not surprising then that this temporal asymmetry is recapitulated in the information processing architecture of a cognitive system that originates in this way. If this information processing is at all relevant to consciousness, then we should expect temporal asymmetry there as well, regardless of whether the universe is in fact an invariant block of space-time, as physics may tell us (in metaphysical parlance: an eternalist universe). From this we have an answer to why the notion that only a dynamic, privileged present exists (in metaphysical parlance: presentism) might be a natural intuition for an organism like us to have.[28] However, when we remember the world as presented to us is not presented *truthfully* but rather *adaptively* and according to our needs (Akins, 1996), the intuition becomes less indicative of the properties of reality and more indicative of properties of ourselves and the way we navigate the world.

Besides explaining perceived temporal asymmetry, the biogenic story can also address perceived dynamism. If we live in a "static" four-dimensional block universe, as relativity suggests, then why does the world seem anything but static? The answer connects the consequences of standard physics just discussed to more recent developments in the sciences of mind. The fundamental dynamism of experience is grounded in the continuous adaptive revision of expectations in light of incoming sensory signals (records) from the environment we are embedded in. This is just what TIER means: Temporality as Iterative Expectation Revision. It should now not seem accidental that we ended up at this particular informational strategy with its particular associated temporal phenomenology (namely, ERA). Instead, this informational strategy and our experience of it should be what we *expect*, given the way the world is and has been.

[28] Huw Price summarizes the intuition well in an interview for the PBS show *Closer to the Truth*: "In the case of time we can be so wrong [about what time really is] because we are highly temporally asymmetric creatures or structures in a region of the universe in which there is a striking temporal asymmetry of the thermodynamic kind. As in the case of up and down, it would simply be a matter of recognition that something we took to be a general feature of the universe is actually just a local feature of our environment," (Price, 2014).

7.8 Conclusion

This chapter's proposal stands out in several ways. The first is that seeing temporality as iterative expectation revision means appealing to the workings of the process, rather than the product, to explain a systemic structural feature of consciousness. TIER also coheres well with an explanatorily powerful larger framework, without bolting on additional theoretical commitments or new mechanisms. Finally, TIER helps to account for the phenomenology in a more elegant and compelling way than has previously been available to proponents of naturalistic theories of consciousness.

In adopting TIER, there are other upshots for our understanding of experience as well. One is an accounting of the subjective arrow of time that can locate the predictive brain within a broader biogenic context, where cognition is dependent upon and conditioned by the physical structure and regularities of the ecological niche we inhabit. There is also a more natural accounting of the extended experiential present than pure content views can offer. This accounting is inclusive of the actual extension of the present experience and the complex content of the present, which seems to include information concerning the organism-relative past and future within the experience.

References

Akins, K. (1996). Of sensory systems and the "Aboutness" of mental states. *The Journal of Philosophy, 93*(7), 337–372.

Bastos, A. M., Usrey, W. M., Adams, R. A., Mangun, G. R., Fries, P., & Friston, K. J. (2012). Canonical microcircuits for predictive coding. *Neuron, 76*(4), 695–711. https://doi.org/10.1016/j.neuron.2012.10.038

Bergson, H. (1889/2001). *Time and free will: An essay on the immediate data of consciousness* (F. L. Pogson, Trans.). Dover Publications.

Brodski, A., Paasch, G.-F., Helbling, S. & Wibral, M. (2015). The faces of predictive coding. *The Journal of Neuroscienc, 35*(24), 8997–9006. https://doi.org/10.1523/jneurosci.1529-14.2015

Brooks, R. A. (1991). Intelligence without representation. *Artificial Intelligence, 47*, 139–159. https://doi.org/10.1016/0004-3702(91)90053-M

Callender, C. (2017). *What makes time special?* Oxford University Press.

Chalmers, D. (1995). Facing up to the hard problem of consciousness. *Journal of Consciousness Studies, 2*(3), 200–219.

Clark, A. (2016). *Surfing uncertainty: Prediction, action, and the embodied mind.* Oxford University Press.

Corcoran, A. W., Pezzulo, G., & Hohwy, J. (2020). From Allostatic agents to counterfactual cognisers: Active inference, biological regulation, and the origins of consciousness. *Biology & Philosophy, 35*, 32. https://doi.org/10.1007/s10539-020-09746-2

Dayan, P., Hinton, G. E., & Neal, R. (1995). The Helmholtz Machine. *Neural Computation, 7*, 889–904.

Dennet, D. (1991). *Consciousness explained.* Little, Brown, and Co.

Dennett, D., & Kinsbourne, M. (1992). Time and the observer: The where and when of consciousness in the brain. *Behavioral and Brain Sciences, 15*(2): 183–247. https://doi.org/10.1017/S0140525X00068229

Dolega, K. (2017). Moderate predictive processing. In T. Metzinger & W. Wiese (Eds.), *Philosophy and predictive processing* (Vol. 10, pp. 161–179). MIND Group.

Doya, K., & Shin, I. (2007). A probability primer. In K. Doya, S. Ishii, A. Pouget, & R. P. N. Rao (Eds.), *Bayesian brain: Probabilistic approaches to neural coding* (pp. 1–13). The MIT Press.

Doya, K., Ishii, S., Pouget, A., & Rao, R. P. N. (Eds.). (2007). *Bayesian brain: Probabilistic approaches to neural coding*. MIT Press.

Engel, A. K., Fries, P., & Singer, W. (2001). Dynamic predictions: Oscillations and synchrony in top-down processing. *Nature Reviews Neuroscience, 2*, 704–716.

Frey, R. D. (1990). Selective attention, event perception and the criterion of acceptability principle: evidence supporting and rejecting the doctrine of prior entry. *Human Movement Science, 9*, 481530.

Friston, K. (2002a). Beyond phrenology: What can neuroimaging tell us about distributed circuitry? *Annual Review of Neuroscience, 25*(1), 221–50. https://doi.org/10.1146/annurev.neuro.25.112701.142846

Friston, K. (2002b). Functional integration and inference in the brain. *Progress in Neurobiology* 68, 113–43. https://doi.org/10.1016/S03010082(02)00076-X

Friston, K. (2003). Learning and inference in the brain. *Neural Networks, 16*(9), 1325–1352. https://doi.org/10.1016/j.neunet.2003.06.005

Friston, K. (2005). A theory of cortical responses. Philosophical *Transactions of the Royal Society B: Biological Sciences, 360*(1456), 815–836.

Friston, K. (2010). The free-energy principle: A unified brain theory? *Nature Reviews Neuroscience, 11*(2), 127–138. https://doi.org/10.1038/nrn2787

Friston, K., & Buzsáki, G. (2016). The functional anatomy of time: What and when in the brain. *Trends in Cognitive Science, 20*(7), 500–511. https://doi.org/10.1016/j.tics.2016.05.001

Gładziejewski, P. (2015). Explaining cognitive phenomena with internal representations: A mechanistic perspective. *Studies in Logic, Grammar and Rhetoric, 40*(1), 63–90. https://doi.org/10.1515/slgr-2015-0004

Gładziejewski, P. (2016). Predictive coding and representationalism. *Synthese, 193*(2), 559–582. https://doi.org/10.1007/s11229-015-0762-9

Godfrey-Smith, P. (2020). *Metazoa*. William Collins.

Grush, R. (2005). Internal models and the construction of time: Generalizing from state estimation to trajectory estimation to address temporal features of perception, including temporal illusions. *Journal of Neural Engineering, 2*(3), S209–S218.

Grush, R. (2006). How to, and how not to, Bridge computational cognitive neuroscience and Husserlian phenomenology of time consciousness. *Synthese, 153*, 417–450.

Grush, R. (2008). Temporal representation and dynamics. *New Ideas in Psychology, 26*(2), 146–157.

Grush, R. (2016). On the temporal character of temporal experience, its scale non-invariance, and its small-scale structure. *Manuscript*. Online at https://doi.org/10.21224/P4WC73: 19.

Hatfield, G. (2002). Perception as unconscious inference. In D. Heyer & R. Mausfield (Eds.), *Perception and the physical world: Psychological and philosophical issues in perception* (pp. 115–144). John Wiley & Sons, Ltd.

Helmholtz, H. (1860/1962). *Handbuch der physiologischen Optik* (vol. 3). Dover.

Hoerl, C., & McCormack, T. (2019). Thinking in and about time: A dual systems perspective on temporal cognition. *Behavioral and Brain Sciences, 42*, e244.

Hohwy, J. (2013). *The predictive mind*. Oxford University Press.

Hohwy, J., Paton, B., & Palmer, C. (2016). Distrusting the present. *Phenomenology and the Cognitive Sciences, 15*, 315–335. https://doi.org/10.1007/s11097-015-9439-6

Hohwy, J., Roepstorff, A., & Friston, K. (2008). Predictive coding explains binocular rivalry: An epistemological review. *Cognition, 108*(3), 687–701.

Husserl, E. (1917/1964). *On the phenomenology of the consciousness of internal time* (1893–1917) (J.S. Churchill. Trans.), M. Heidegger (Ed.). Indiana University Press.

Ismael, J. (2021). *Time: A short introduction*. Oxford University Press.

James, W. (1890/1980). *The principles of psychology*. Henry Holt.

Kent, L., & Wittmann, M. (2021). Time consciousness science and its theories. *Neuroscience of Consciousness, 7*(2), niab011. https://doi.org/10.1093/niab011

Kiebel, S. J., Daunizeau, J., & Friston, K. J. (2008). A hierarchy of time-scales and the brain. *PLoS Computational Biology, 4*(11), e1000209. https://doi.org/10.1371/journal.pcbi.1000209

Lyon, P. (2006). The biogenic approach to cognition. *Cognitive Processing, 7*(1), 11–29. https://doi.org/10.1007/s10339-005-0016-8

Malapani, C., Deweer, B., & Gibbon, J. (2002). Separating storage from retrieval dysfunction of temporal memory in Parkinson's disease. *Journal of Cognitive Neuroscience, 14*, 311–322.

Mathys, C., et al. (2012). Computational modeling of perceptual inference: A hierarchical Bayesian approach that allows for individual and contextual differences in weighting of input. *International Journal of Psychophysiology, 85*(3), 317–318

Mattes, S., & Ulrich, R. (1998). Directed attention prolongs the perceived duration of a brief stimulus. *Attention, Perception, & Psychophysics, 60*, 1305–1317. https://doi.org/10.3758/bf03207993

McTaggart, J. M. E. (1908). The unreality of time. *Mind, 17*(68), 457–474.

Merleau-Ponty, M. (1945). *Phenomenology of perception* (Colin Smith. Trans.). Routledge.

Montemayor, C. (2013). *Minding time: A philosophical and theoretical approach to the psychology of time*. Brill.

Montemayor, C., & Wittmann, M. (2021). The illusions of time passage: Why time passage is real. *Preprint*. https://doi.org/10.31234/osf.io/zn3s6

Oettinger, A. G. (1966). The uses of computers in science. *Scientific American, 215*(3), 160–175.

Parr, T., Pezzulo, G., & Friston, K. (2022). *Active inference: The free energy principle in mind, brain, and behavior*. MIT Press.

Pennartz, C. M. A. (2015). *The brain's representational power*. MIT Press.

Price, H. (1996). *Time's arrow and archimedes' point: New directions for the physics of time*. Oxford University Press.

Price, H. (2014). Interviewed by Robert Kuhn for *Closer to the Truth*, PBS/Youtube, August 25. Available at Retrieved April 15, 2022, from https://www.youtube.com/watch?v=1_C8UzwQ63g

Rao, R. P., & Ballard, D. H. (1999). Predictive coding in the visual cortex: A functional interpretation of some extra-classical receptive-field effects. *Nature Neuroscience, 2*, 79–87.

Reynolds, J. R., Zacks, J. M., & Braver, T. S. (2007). A computational model of event segmentation from perceptual prediction. *Cognitive Science, 31*(4), 613–643. https://doi.org/10.1080/15326900701399913

Roseboom, W., Fountas, Z., Nikiforou, K., et al. (2019). Activity in perceptual classification networks as a basis for human subjective time perception. *Nature Communications, 10*(1), 267.

Russell, B. (1924/1985). Logical atomism. In D. F. Pears (Ed.), *The philosophy of logical atomism* (pp. 157–181). Open Court.

Salti, M., Harel, A., & Martin, S. (2019). Conscious perception: Time for an update? *Journal of Cognitive Neuroscience, 31*(1), 1–7. https://doi.org/10.1162/jocn_a_01343. Epub 2018 Sep 21.

Seth, A. (2021). *Being you: A new science of consciousness*. Faber and Faber.

Shapiro, F. R., & Henderson, J. R. (2006). *The yale book of quotations*. Yale University Press.

Sims, M. (2021). A continuum of intentionality: Linking the biogenic and anthropogenic approaches to cognition. *Biology & Philosophy, 36*, 51.

Solms, M. (2021). *The hidden spring: A journey to the source of consciousness*. Profile Books.

Swanson, L. R. (2016). The predictive processing paradigm has roots in Kant. *Frontiers in Systems Neuroscience, 10*, 79. https://doi.org/10.3389/fnsys.2016.00079

Thompson, E. (2010) *Mind in life: Biology, phenomenology, and the sciences of mind*. Harvard University Press.

Tversky, B., & Zacks, J. M. (2013). Event perception. In D. Reisberg (Ed.), *Oxford handbook of cognitive psychology* (pp. 83–94). Oxford University Press

Wacongne, C., Labyt, E., van Wassenhove, V., Bekinschtein, T., Naccache, L., & Dehaene, S. (2011). Evidence for a hierarchy of predictions and prediction errors in human cortex. *prediction errors in human cortex. Proceedings of the National Academy of Sciences (PNAS) USA, 108*(51), 20754–20759. https://doi.org/10.1073/pnas.1117807108

Wiese, W. (2017). Predictive processing and the phenomenology of time consciousness: A hierarchical extension of Rick Grush's trajectory estimation model. In T. Metzinger & W. Wiese (Eds.), *Philosophy and predictive processing* (26). MIND Group. https://doi.org/10.15502/9783958573277

Wiese, W., & Metzinger, M. (2017). Vanilla PP for philosophers: A primer on predictive processing. In T. Metzinger & W. Wiese (Eds.), *Philosophy and predictive processing* (p. 1). MIND Group. https://doi.org/10.15502/9783958573277

Wolfe, J. M. (1983). Influence of spatial frequency, luminance, and duration on binocular rivalry and abnormal fusion of briefly presented dichoptic stimuli. *Perception, 12*(4), 447–456.

Zacks, J. M., Speer, N. K., Swallow, K. M., Braver, T. S., & Reynolds, J. R. (2007). Event perception: A mind-brain perspective. *Psychological Bulletin, 133*(2), 273. https://doi.org/10.1037/0033-2909.133.2.273

Chapter 8
Conclusion: Looking Back and Looking Forward

8.1 Looking Back

This book has proposed a new way of thinking about temporal experience that operates across multiple levels of analysis to provide a coherent and plausible understanding of time-consciousness. This view involves several distinctive "stances." The first stance is the idea that time, for the subject, should be reconceptualized as something quite different and apart from both objective time and any individual sensory capacity we might have. Instead, I have contended that time-consciousness is *special* in the following ways: (1) it is the precondition for subjective, agentive experience as we know it; (2) it is not something we "sense" but rather something we construct through the ongoing activity that creates conscious experience; and (3) because it is constructed, subjective time should properly be thought of as the *way* or *manner* by which we experience (including perception but also, imagination, memory, etc.). This third point, which I am calling adverbialism about subjective time, represents an overall orientation of the present work. I think that, upon reflection, the adverbialist stance should turn out not to be very controversial, and its primary competitor, viz. the notion that we—and the brain—sense objective time directly,[1] is symptomatic of ignoring what is given in phenomenology as well as falling prey to seductive analogies from well-studied kinds of sensory perception like vision.

Besides these general stances, this book has defended the "doctrine of the specious present," which contends the experienced present occupies an interval, from attack by snapshot theories of various sorts, and has argued that this specious present should really be taken as a further explanandum. I then argued in favor of a pluralistic explanatory strategy towards temporal phenomenology, the upshot of

[1] I have said little about this view directly, as it is not often explicitly defended. The internal clock models discussed in Chap. 7 are probably the most prominent examples of this way of thinking.

C. A. McKenna, *Constructing the Present*, Studies in Brain and Mind 24,
https://doi.org/10.1007/978-3-031-89571-5_8

which I have called ERA, short for Extensional-Retentional Analysis. This view considers the real temporal extension of cognitive and physiological processes to be explanatorily relevant to the character of our current experience, while at the same time suggesting that which can be usefully interpreted as intentional content is also important to the richness of time-consciousness. More precisely, I have argued we should appeal to properties of a real temporal extension of experience in order to ground the phenomenology of the specious present and the fundamental fact of temporal passage, while we may also appeal to the properties of intentions when explicating certain aspects of our phenomenology, such as the temporal relations of events. The relevant properties of the real temporal extension of experience for time-consciousness come in two important forms: the ever-changing relations of the subject to experiential objects, which provides for experiential succession, and the duration of experience that these changing relations occur and are integrated over, which grounds the finite and variable duration of the specious present. I have argued this model not only better accommodates the phenomenology of time-consciousness than its rivals, but also dovetails with the predictive approach to the mind through the framework of Temporality as Iterative Expectation Revision (TIER). Linking the phenomenological analysis to a particular view about the cognitive processing provides specificity and potential empirical tractability, while also inviting a synthesis of levels of analysis that has until now remained elusive. Ultimately, it is my hope that we may further fill in the gaps of how this phenomenological analysis ought to be implemented cognitively and, eventually, physiologically on the level of neural implementation.

8.2 Looking Forward

The study of time-consciousness remains in its infancy, both philosophically and empirically. As such, I believe the final answers to most of the interesting questions in this domain will continue to evade capture for quite some time. However, it is my hope that the various proposals within this book turn out to have significant explanatory value and practical utility in the context of future research directions while possibly also representing an advance in our understanding of the phenomenology of time-consciousness and its underlying mechanisms. Naturally, more work remains to be done to establish whether this is the case.

Whatever our best theory or theories of consciousness and cognition turn out to be, it seems to me that these should determine more than anything the sort of theory of time-consciousness we endorse. Such an overarching theory can both create constraints on theories of time-consciousness and point towards the role that time must play in consciousness. I have therefore attempted to elaborate upon what a theory of time-consciousness might look like according to broadly predictive theories of mind. Of course, the predictive approach to the mind might well turn out to be misguided. But if it is not, then investigating the workings of this cognitive apparatus would appear to be the best way to illuminate the place of time-consciousness in the overall system.

It is also important that we endeavor to know more about the brain than we currently do. Philosophers can exercise themselves to their hearts' content about what is logically possible and what is not, but it should be obvious that what the brain is actually capable of and, moreover, what it is *actually doing* will always be what really matters. While *a priori* argument underdetermines the position we should take in this domain, introspection, even with great attention paid to phenomenological detail, is notably fallible. It may not seem to the layperson, for instance, that they experience anything other than a momentary slice of time, and yet the orthodox view holds that is not how the mind works in practice. The advent and continual corroboration of the specious present is therefore an advancement over the pre-theoretical understanding of time-consciousness. Other advancements like this are possible and may provide greater insight into the fine-grained character of our own temporal phenomenology. For example, we could certainly benefit from knowing more about what is happening during postdictive effects on the neurological level. A greater understanding of this topic could lead to a revision of our interpretation of these strange phenomena but also to further insights about time-consciousness under normal conditions.

One area of relevant current research that has been attracting a growing amount of attention lately concerns the neural correlates of time-consciousness. These are the supposed physical substrates of our experience of the present. So far, accounts of these neural correlates have remained rather speculative. While there have been scientific attempts to locate the neural correlates of time-consciousness, results have been inconclusive. In general, mainstream psychology has attempted to address the question of the neural correlates of time-consciousness in a different way than theories arising from a cognitive science perspective (Lloyd, 2012). For instance, while it remains common to speak of physiologically instantiated "internal clocks" in the psychological literature (Wearden, 2016; Allman et al., 2014), cognitive (neuro) scientists like Anil Seth (2021) are confident there is "no dedicated sensory system for time" and "there is no evidence for any 'neuronal clock' inside the head which measures out our experiences in time" (Seth, 2021, p. 134; see also Wittmann & Montemayor, 2022, p. 362). Instead, as previously discussed in Chap. 7, Seth endorses recent work by Roseboom et al. (2019) that temporal experience, understood as "duration perception," is the result of inferences about the rate of perceptual change—a process which has been modeled in an artificial neural network (ibid.).

There is thus a division in the scientific literature between those that search for a particular physiological mechanism underlying time-consciousness and those that appeal to particular mechanisms within some overarching information-processing framework. As we have seen, examples of the latter strategy, besides Seth and Roseboom, include Rick Grush's "Trajectory Estimation Estimation Model" (2005, 2006) and Wanja Wiese's (2017) "hierarchical extension" of Grush's theory, which situates that computational model within the "predictive processing" framework. Others, such as Hohwy et al. (2016), have attempted to give predictive accounts of the "flow of time" that appeal to the cognitive system's higher-order expectations of change. The proposal in Chap. 7, that temporality is grounded in iterative expectation revision in the context of the mind as an ever-active prediction generator, is also of this kind.

A further sort of endeavor attempts to ground temporal phenomenology in neurophysiological processes, with less emphasis on the computational-level information processing. For example, Francisco Varela (1999, 2000), Varela et al. (2001) sees time-consciousness as a globally emergent phenomenon, locating its NCCs in large-scale patterns of neural oscillations. Georg Northoff (2016, 2023) likewise seeks answers from brainwaves, though he is keen to distinguish "neural predispositions of consciousness" (NPCs) from neural correlates. Dan Lloyd directly addresses the "neural correlates of temporality" by advancing the idea that the default mode network (a global brain state) plays a key role. Lloyd rightly recognizes that, unlike the "timing" paradigm common to psychological studies, the more important subject of investigation is temporality as "a structural feature of consciousness" and "a fundamental dimension of all percepts and all behavior" (Lloyd, 2012, p. 695). For Lloyd, the default mode network is meant to underpin temporal phenomenology as it is characterized by James and Husserl. Lloyd justifies his conclusion by appealing to an apparent mapping between the temporal characteristics of the default mode network and the temporal structure experience. However, as Grush (2006), referring to an earlier version of Lloyd's view, and Lee (2012) have argued, these types of proposals remain conceptually and methodologically problematic, even if the background assumptions are correct.

The main problem with low-level implementational views is not so much that they are necessarily off track but, given the state of knowledge right now, they can appear correlational and arbitrary without a tight explanatory connection with cognitive processing. While it is perfectly reasonable to expect experience to be grounded in physical processes, such explanations need some kind of plausible account of why the proposed physical features in particular should be determinative of temporal phenomenology, and this explanation should be connected to what the brain is actually doing to generate consciousness. As our understanding of the latter remains extremely poor, so too does our ability to provide such explanations. However, the properties of the neurophysiological processes crucially involved in the construction of subjective experience would naturally seem most likely to bear on the temporal quality of the resultant experience, at least more so than a temporal property found to vaguely correlate to modulations in subjective experience in an fMRI machine. Even then, we cannot assume that the connection between those relevant physical properties and the experience is anything like straightforward and may be much less illuminating than an investigation at a higher level of abstraction.

That being said, Varela and Lloyd's suggestions for the neural correlates of time-consciousness, though not without issue, are most in the spirit of the current project because they contend that structural features of more general neural correlates of consciousness are responsible for the temporal character of experience, rather than the temporal character of experience resulting from *additional* physiological or computation processes. Likewise, Montemayor and Wittmann have recently suggested an NCC for "time passage" at the information processing level which is not separate from the general model updating procedure that a conscious system engages in to be conscious in the first place (Wittmann & Montemayor, 2022). These perspectives are more promising than either the classical approach, which searches for specific physiological timing mechanisms, or approaches from cognitive science that explain time-consciousness through "bolt-on" computational modules to some favored information processing mechanism.

It is my hypothesis that any attempt to identify a specific "location" or single network responsible for experienced time will continue to fail. However, with a promising overall theoretical framework, it may be that important aspects of relevant brain processes might be identified, and possibly even intervened upon to corroborate what I have suggested about the critical importance of time-consciousness to subjectivity as a whole. As is well known, there is a vast array of conditions that act as temporal modulators, and while psychology has done a great job identifying what these are (viz., seemingly anything you can think of), we are still very much in the dark about *why* such conditions act as temporal modulators, or why some have the effects they do while others have the opposite effect. Greater insight into the processing underlying the effects of these conditions (e.g., pharmacological intervention, boredom, attention, and so forth) might point us further in the right direction concerning the specific mechanisms underlying time-consciousness.

Another potential avenue for further progress is to approach time-consciousness *biogenically*. From the biogenic perspective, consciousness and cognition form a continuum with life, and thus, looking towards deep evolutionary history may be one place to gather further clues about the character and genesis of human time-consciousness. While the penultimate chapter gives a hint of how a biogenic story can inform these kinds of accounts, there is certainly more that can be learned about time-consciousness from studying the origins of life and cognition in a variety of living systems. In combination with empirical work on neural correlates of time-consciousness, there may be synergies that lead us to a different understanding of the adaptive import of time-consciousness and the possible permutations of temporal experience in non-human animals, which may, in turn, shed further light on our circumstances.

Over the course of these pages, I have advanced a novel view of time-consciousness from multiple perspectives and at various levels of abstraction and analysis. This view provides a coherent alternative to current theories treating this subject. In a nutshell the proposal is this: time-consciousness is not simply an aspect of experience but rather a precondition. It is the way experience must happen. This thought motivates a more detailed phenomenological analysis that sees the actual extension of experience in time and the contents thereof as both explanatorily relevant to the ultimate temporal character of experience. This thought has also motivated the corresponding information-processing picture of temporality as iterative expectation revision, which was initially inspired by the advent of the predictive approach to the mind. On this view, we construct the present in the very activity of being conscious. If this is right, then it turns out that time is not an object of experience at all, but rather how we experience the world.

References

Allman, M. J., Teki, S., Griffiths, T. D., & Meck, W. H. (2014). Properties of the internal clock: First- and second-order principles of subjective time. *Annual Review of Psychology, 65*, 743–771.

Grush, R. (2005). Internal models and the construction of time: Generalizing from state estimation to trajectory estimation to address temporal features of perception, including temporal illusions. *Journal of Neural Engineering, 2*(3), S209–S218.

Grush, R. (2006). How to, and how not to, Bridge computational cognitive neuroscience and Husserlian phenomenology of time consciousness. *Synthese, 153*, 417–450.

Hohwy, J., Paton, B., & Palmer, C. (2016). Distrusting the present. *Phenomenology and the Cognitive Sciences, 15*, 315–335. https://doi.org/10.1007/s11097-015-9439-6

Lee, G. (2012). Commentary on Dan Lloyd: "Neural correlates of Temporality". *Consciousness and Cognition, 21*(2), 707–710. https://doi.org/10.1016/j.concog.2011.06.012

Lloyd, D. (2012). Neural correlates of temporality: Default mode variability and temporal awareness. *Consciousness and Cognition, 21*, 695–703. https://doi.org/10.1016/j.concog.2011.02.016

Northoff, G. (2016). Slow cortical potentials and 'inner time consciousness'—A neuro-phenomenal hypothesis about the 'width of present'. *International Journal of Psychophysiology, 103*, 174–184.

Northoff, G. (2023). *Neurowaves: Brain, time, and consciousness*. McGill-Queen's University Press.

Roseboom, W., Fountas, Z., Nikiforou, K., et al. (2019). Activity in perceptual classification networks as a basis for human subjective time perception. *Nature Communications, 10*(1), 267.

Seth, A. (2021). *Being you: A new science of consciousness*. Faber and Faber.

Varela, F. (1999). Present-time consciousness. *Journal of Consciousness Studies, 6*(2–3), 111–140.

Varela, F. (2000). The specious present: A neurophenomenology of time consciousness. In J. Petitot, F. J. Varela, B. Pachoud, & J.-M. Roy (Eds.), *Naturalizing phenomenology: Issues in contemporary phenomenology and cognitive science* (pp. 266–314). Stanford University Press.

Varela, F. J., Lachaux, J.-P., Rodriguez, E., & Martinerie, J. (2001). The Brainweb: Phase synchronization and large-scale integration. *Nature Reviews Neuroscience, 2*, 229–239.

Wearden, J. (2016). *The psychology of time perception*. Palgrave Macmillan.

Wiese, W. (2017). Predictive processing and the phenomenology of time consciousness: A hierarchical extension of Rick Grush's trajectory estimation model. In T. Metzinger & W. Wiese (Eds.), *Philosophy and predictive processing* (Vol. 26). MIND Group. https://doi.org/10.15502/9783958573277

Wittmann, M., & Montemayor, C. (2022). Reinterpreting the Einstein-Bergson debate through contemporary neuroscience. In S. Gozzano (Ed.), *Einstein vs. Bergson: An Enduring Quarrel on Time*. De Gruyter.

Index

C. A. McKenna, *Constructing the Present*, Studies in Brain and Mind 24,
https://doi.org/10.1007/978-3-031-89571-5

The manufacturer's authorised representative in the EU is Springer Nature Customer Service Centre GmbH, Europaplatz 3, 69115 Heidelberg, Germany. If you have any concerns regarding our products, please contact ProductSafety@springernature.com

Printed and bound by CPI Group (UK) Ltd, Croydon, CR0 4YY
30/09/2026
02225570-0002